Esther Woolfson was brought up in Glasgow and studied Chinese at the Hebrew University of Jerusalem and Edinburgh University. Her acclaimed short stories have appeared in many anthologies and have been read on Radio 4. She is the author of *Corvus: A Life with Birds*, which is also published by Granta Books. She has won prizes for her nature writing and received a Scottish Arts Council Travel Grant and a Writer's Bursary.

'The joy of this book is that it teaches us to observe afresh and to live peaceably with every species – no matter how slimy' *Daily Telegraph*

'The book gives us a picture of wildlife in Aberdeen, a city not remarkable for the richness of its wildlife except for its gulls; but that doesn't matter, because what Woolfson specialises in is demonstrating how little most people understand about creatures they think ordinary. As is only to be expected from the author of that delightful book *Corvus*, she does this well ... Her book amounts to a series of essays with groups of shorter pieces in between them, and the essays are on subjects very much to her taste, so her writing flows beautifully' Diana Athill, *Literary Review*

'Woolfson deserves only praise for bringing her hidden city – and our hidden prejudices – so enjoyably to light' *Sunday Telegraph*

'Woolfson changed our minds about rooks and crows with her memoir *Corvus* and now her new book is going to change our minds about other species, specifically those that have moved into the cities and made their home among the glass and the concrete' *Sunday Herald*

'*Field Notes from a Hidden City* is a measured, lyrical and idiosyncratic chronicle of a year in Aberdeen – but more than that, it's a plea to value the everyday richness we have on our doorsteps, and to mobilise against its loss: not, perhaps, by campaigning (though that can surely be useful), but by noticing, and respecting, and perhaps daring to believe for one wild moment that our human needs are not the only ones that count' *Caught by the River*

'A wonderful writer … with this expansive, thoughtful and profound book Woolfson has added another tome to the canon of wonderful modern nature writing' *Big Issue*

'A beautiful and absorbing account of the natural world' *Liverpool Post*

'*Field Notes from a Hidden City* is highly engaging; learned, thoughtful, studded with anecdote, aphorism, myth and scientific facts drawn from literature in its widest sense' *Times Literary Supplement*

'Woolfson's quest is to record her encounters with urban wildlife over the course of a year, from a young pigeon she rescues from the snow to the spiders she finds in her house … in precise, and often lovely, prose' *Financial Times*

'This is a quietly enthralling, generous book; Woolfson's sympathy and precision of observation makes new observers of her readers too' *Civilian*

FIELD NOTES FROM A HIDDEN CITY

Esther Woolfson

GRANTA

Granta Publications, 12 Addison Avenue, London W11 4QR

First published in Great Britain by Granta Books, 2013
This paperback edition published by Granta Books, 2014

A CIP catalogue record for this book
is available from the British Library.

3 5 7 9 10 8 6 4 2

ISBN 978 1 84708 276 3

Typeset by M Rules
Printed and bound by CPI Group (UK) Ltd, Croydon, CR0 4YY

MIX
Paper from
responsible sources
FSC® C013604
www.fsc.org

In memory of my father, and of a great musician,
my cousin Eric Woolfson (1945–2009)

... we would assume that what it is we meant

would have been listed in some book set down
beyond the sky's far reaches, if at all
there was some purpose here. But now I think
the purpose lives in us all and that we fall

into an error if we do not keep
our own true notebook of the way we came,
how the sleet stung, or how a wandering bird
cried at the window ...

Loren Eiseley

Contents

Introduction   1

Snow   11
  November 24th to December 20th   13

Midwinter   35
  December 21st to February 19th   37
  Fellow Travellers   62

Winter into Spring   81
  March 1st to March 9th   83

Early Spring   103
  March 14th to March 25th   105
  Orb-weavers and Wanderers   121

Spring   141
  April 5th to April 15th   143

CONTENTS

Late Spring   161
  April 18th   163
  Angels in the Streets   175

Spring to Summer   193
  April 26th to May 10th   195
  The Unlikely Cupid   206

Early Summer   219
  May 23rd to July 11th   221
  The Bird with the Silver Eyes   245

Midsummer   257
  July 17th to July 26th   259
  Flying through the Storm   268
  July 28th   283
  The Fugitive in the Garden   288

Into Autumn   309
  August 10th to November 2nd   311

Acknowledgements   344
Select Bibliography   346
Text Credits   349
Index   351

INTRODUCTION

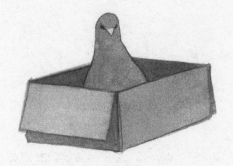

It was almost four in the afternoon on one of the oddly quiet days of December. Snow had fallen again, yet another layer to freeze onto the iron-hard strata of thick, packed ice. In less than a month, the city had been turned into a fortress of ice and stone. It was already dark as I walked past the church at the end of the lane. A line of bright lights from house windows seemed to catch at the sparkle in the air. Every afternoon towards dusk, the cold became visible; it fell in a fine mist of ice particles which stung in the throat. By the time it was dark, it had glazed every surface to an even glitter.

Christmas soon, and a year ending. The year had been one of superlatives – the lowest recorded temperatures, the heaviest snow, the wettest summer; in matters economic as well as meteorological, of no less magnitude and equally unassailable. It seemed a year designed to test every certainty, to bring the stirrings of disquiet, to make you wonder what might be next and when.

Illuminated by a streetlight, something in the lane moved in the snow, a frantic patch of thrashing darkness, a flail and panic of blue and grey. Whatever it was, I ran clumsily towards it, impeded by paddings of clothes, socks and boots. It was a bird, almost submerged, drowning in snow. I scooped it up from the drift in both hands; a young pigeon, still small with the pink, moist-looking beak of a recent fledgling. He was feathered but his feathers were drenched and he carried one wing awkwardly. There were no other birds around. It was past roosting time, too late and too cold for them. I folded the wing carefully against the bird's body and held him. I'd been about to go to the corner shop but suspected he wouldn't be welcome there. There was nothing I could do but carry him home, beyond immediate danger. It was only as I've done with generations of the avian needy before him.

As we walked, the bird peered from between my hands, a feral pigeon, a blue rock pigeon, *Columba livia*: future street-dweller, serial pavement-picker, underfoot obstacle, as familiar and ubiquitous as ourselves, one of the latest generation of the birds who resolutely populate the cities of the world, a parallel presence to our own, knowing, able, urban survivors, accepted and decried in varying measures. I could feel his fast, panicky heartbeat, even through thick gloves.

At home, I wrapped the bird in a towel to dry him, carrying him swaddled under my arm as I lined a box with newspaper. When he was dry, I put him in and he settled immediately. I put the box in the vestibule, where it was just warm enough – the bird had his full plumage, and now that he was dry, he'd be able to keep himself warm. Later, I'd think about all the possible problems of raising him but first I had to go and finish my shopping.

Walking back down the lane, past garden walls tessellated by blown snow, I realised that the bird's parents must have been roosting among the stone buttresses of the church near to where I found him. Pigeons breed at unlikely times although that moment seemed even less judicious than most. I wondered how he'd come to be there – fallen, pushed, lost. With birds, you never really know.

On my way back home again, I stopped a moment. The place where he'd been struggling was now a patch of compacted snow, ridged and imprinted with the broad zigzag indentations that must have been made by the tyres of the Range Rover that had driven slowly past us in the lane as we walked home together.

Over the next few days, the bird stayed contentedly enough in his box. I would wait until the wing was fully healed and then let him go, return him to the wild.

As I fed and took care of the bird, I began to think more about the expression 'return to the wild', the words we use to describe setting a creature free to return to its own environment, to live its life among its own, a natural life. The wild in this case would most probably be the large sandstone church at the end of the lane, a neo-Gothic building of step-corbelling and intricate stone finials, around whose tall central spire crows and magpies engage in glorious aerial combat on windy days. It would be the roofs and gardens of this district, the handsome Victorian villas, long-established trees and wide, quiet streets only a short walk from the centre of the city. Hardly wild.

For all that, this was a wild bird – a wild, city bird although the words 'wild' and 'city' seemed difficult to reconcile. I began to think about wildness in relation to creatures who live in cities, about whether or not we consider them less wild than creatures living elsewhere, or think of them as somehow a lesser part of nature itself. I wondered if the same might apply to humans, as if merely by being in a city, not only might our lungs be polluted but ourselves, our minds (and if we have them, souls), as if urban dwellers must by definition be over-avid consumers of the unnecessary, weakened by purchase, alienated in every way, distanced from a lost, admonitory Eden. Are we, I wondered, living lives remote from all that is natural, beneficial, wild, or are we as much a part of the natural ordering of the universe as the wildest of things, moved by the same forces, as wild as anything else on earth? It seemed to me that if I'm representative of my species, the bird was representative of many more, of all the other birds, beasts and smaller life forms, seen and unseen, who populate our cities, who make their lives among us – and were often there before us – adapting to our habits, sharing with us the triumphs or follies of our building policies, our tendencies to destroy and pollute or to engage in the ecologically catastrophic. Their presence may be the only contact many urban people have with the natural world but our relationship with them seems changed by proximity, diminished by the very fact of their being here among us. As I'd passed the compacted snow where I'd picked up the bird, I had begun to think of the value of any creature's life and of the ways in which we make judgements or calculate worth, of the complex gradations we apply to other species, the ones we use to approve or to condemn.

I passed the bird on the way in and out of the house, checking on him often to make sure that he was eating and roosting. In the mornings, I went downstairs with the usual trepidation, the kind you always have when you're looking after a rescued bird when you wonder if some unknown disease or frailty has ended its life in the night. But every morning he was there, alive.

A wild bird but an ordinary one. I looked up the definition of 'ordinary' – *With no special or distinctive features, common, of ordinary rank, undistinguished, commonplace.* Which of us is any more or less? Whatever he was, this bird was beautiful. His new, fresh feathers were lavender and navy, shading to a fine line of black towards the tips of his wings, his eyes bright and watching.

In that dark and quiet city lane as we'd walked past the backs of offices and car parks, empty and deserted on that late Saturday afternoon, the bird had seemed symbolic of something, although I wasn't sure what. It may have been the season, the approach of Christmas, which made me think of things abandoned and needy. Or perhaps it was just the depictions of doves which are, after all, pigeons by another name, in attitudes of unlikely spiritual bliss that were everywhere on cards and decorations. It may have been the winter, the worst many people had ever experienced which, in its startling and prolonged severity, had brought an atmosphere of strangeness, a feeling of closing in and uncertainty, that had encouraged me to feel reflective. This creature seemed to symbolise the fragility that suddenly I felt was there, at the heart of everything.

The fierce and early cold had already made me hyper-aware of our place on the surface of the earth, forcing on me a new kind of realisation about the effects of weather, the ways in which we all live with it in precarious dependence. And now the bird was making me look differently at the life of the city. As I watched him, I wondered what we were doing here together. How had we come to be in this place, at this moment? What circumstances had brought us to this city? What allowed and even encouraged me to pick up this member of another species? What was it that made me care? What made him accept my human attentions without fear or panic? Any relationship between us, our individual lives and fortunes too, felt both momentous and quotidian, the way most of life turns out to be. Together, in spite of our obvious differences, we were as bound as any two creatures on earth by something only too measurable – life itself. Living in a city, we are all elements of a biological and ecological chain described by words that express the complex web of connection between us and hint of dependency and need – *commensal, mutual, symbiotic, predatory, synanthropic.* Our streets, buildings, houses are shared, our gardens and our trees. In one city, there are more cities than we know, hidden cities inhabited by those with whom we share everything we rely on: food and light and air. In differing degrees, we share our vulnerability to the elements that shape and dominate our lives: cold or heat, wind and rain. If it was true of this bird, it's true of everything else that lives in any city; any creature, the ones too we bring into our lives by choice.

During the first snow, the quality of those strange, cold days seemed to require something of me. It was the necessity to record, to keep an impression of the time in some small way, a way beyond forgetting. Finding the bird only concentrated the feeling, and I knew that if I should document our lives and time and place, I should do it now. As the bird watched me intently from his box, it felt as if it was the least that I might do.

Over the days, the damaged wing improved. The bird grew bolder, stood on the edge of his box, became impatient. He must have flown because one morning, I had to pick him out of the leaves of the plants on the recessed windowsill, then from the high ledge above the door.

I hoped that he would be ready to release before Christmas. My family would be coming home, and there'd be enough to do. I'd have to decide when to let him go and hoped that his strong homing sense, that powerful, complex set of navigational tools pigeons have, would be intact. But could he navigate in snow? Was he old enough to find his way?

On a morning a few days before Christmas, I opened the door to see how much more snow had to be shovelled from the front path. It was only a little open but it was enough. The bird flew swiftly onto the top of the door, removing from me any need for decision. Quickly he was off, seeming to turn right although I couldn't be quite sure if it was really him I saw in the early-morning light, heading towards the lane, towards the church, out into the wild and snowy city.

SNOW

November 24th

The first snow of winter begins to fall in the afternoon, while it's still light. It lies at once although it's too early for it. Usually, the first thin flurries of the season fall quickly then melt away but today, it is a chill more intense than most November cold. Dark before four, and still snowing. Light is short now – at the Winter Solstice, less than a month away, there'll be only six hours and forty minutes of light from dawn until dusk. The light clicks away from the Summer Solstice to the Winter, minute by minute into darkness, and then slowly back again to light. By early evening, the snow is already thick enough to muffle the sound of rush-hour traffic.

Last winter seems only recently over. Its snow didn't melt fully until late April and it snowed again in June. Even when it began to thaw, the process seemed endless, slicks of glazed white clinging for weeks to the foot of north walls and filling hollow dents under trees. With it, there seemed to be a gradual sense of recovery as if we all had to cast off the shadow of cold before we could move into the brightness of spring and summer, a brightness that never came. Quick blusters of flying flakes belaboured the days of May, melting before they lay. Like an echo of snow, crack-shots of hail blew in volleys against glass and slate in June. Cold winds tore down branches and scattered the ground with fallen petals of rose and rhododendron. During July and August, the granite from which this city is built was dark with water. Even in good years, days of summer sun are scarce – any of the hedonistic impulses you might feel if you lived in a more encouraging climate ebb away with the season and dissolve into chill. (How fortunate we are – we need have no regrets at the ending of hot and glorious summers.) The year's snow and rain foreshortened time, seemed to press it into less of a measure than it should take, into an altered, flattened dimension. The seasons themselves felt bent into unrecognisable shapes, lopped or stretched, lasting only days, flaring then dying back. The transition from what might have been spring and then summer into what might have been autumn was imperceptible – the name of the month, from rain to rain.

For a while, I thought I was the only one to feel disorientated by this disjunction of seasons and time but other people were too. Friends reported feeling Novemberish in July, expectant but disappointed by the time summer came, as if spring hadn't yet been. Our progress through the months seemed to have happened in a muted circle, unrelated to anywhere else. Other European cities with burning summer temperatures might have been in a different zone of time, on an entirely different continent. We had six months, more or less, of snow.

There was even more discussion than there usually is of weather and natural phenomena, of what might be concluded from one year's measuring, of what's measurable and what's not. The weather and our discussion felt unprecedented, as if both were just part of an infinitely larger picture, one whose totality we couldn't yet see. In the months between the snow, there had been ash clouds and rain; in other places, extreme heat and earthquakes, oil spills and drastic flooding. And now this new snow, which feels unexpectedly like a completion, the closing of a circle.

During the year I had a desire to travel, but now it has disappeared completely, almost as if it has dissolved under the weight of snow. I hadn't been anywhere for a long time – anywhere far away – and it made me feel restless. I began to plan and to think about where I wanted to go. It was only a modest, unambitious plan that would have taken me to one of the European cities for a few days, but even that kept being overturned by one thing or another, first by the weather and then work and after that, airline crises. I put it into abeyance and almost forgot about it but now, I remember it again. The restlessness has vanished and I'm delighted not to have to consider being anywhere else at all. (It's just as well because the trains have stopped and the airport is closed and all but the main roads are impassable.) It's a world in miniature. The only way I have to understand it is within a different set of coordinates, to see it in close-up, to know what's near.

November 25th

The second morning, and already it seems like time marked out and possibly named: The Snow. I have to drive a little way out of town on as yet ungritted roads for a gathering of friends who don't see each other often but instead of ten, we're three. Most of the others sensibly won't drive in this weather from Perthshire and Angus. We sit in a large conservatory and talk about work and books and watch more snow fall.

Although there's no reason yet to think it, the others too feel that this is just the beginning. We discuss stocking up, preparing.

On the way home, I stop to buy some of the things I might not be able to carry easily if this weather carries on: kindling and logs, bags of flour, wheat and rye and spelt, as if I'm preparing for a Russian winter.

When we came here first over twenty years ago, the weather was different, in memory at least. I was familiar with the west of Scotland, with relentless rain, seasons of rain, rain-lashed autumns and winters. For me, east-coast weather meant the ferocious winds of Edinburgh. By comparison, the north-east seemed chill and bright and dry, its snow measured, limited to a few days in the first months of the year and then over. The first January we were here, I bought new sunglasses against the glare of low winter sun.

(I take into account the nature of perception and the unreliability of memory. I know the retrospective, nostalgic tricks the mind uses in observation of the past. Have there really been more storms, more rain, unexpectedly intense moments of heat, followed by quick reverses into cold?)

November 27th

A few days now, and the snowfall has settled into a pattern. Every day, shifting, vertical blinds of moving grey and white drop from pale violet or charcoal skies. Each fall hardens to glittering ice in the freezing, crystalline air of night and then every morning, more snow falls. Even though it has only been for a short time now, I begin to remember how prolonged snow and darkness make life seem small. Already, everything seems to have shrunk to the possible or just the necessary. The cold seems to seal us all in. I can't decide if we're cocooned within the city, in our houses, or isolated inside ourselves. Any of these seems fine to me. These days of winter are brief and freezing and often brilliant and then, all the rooms are filled fleetingly with a pure white light.

Cold and darkness alter everything; they change the pattern of days and nights, make me question what is necessary. Towards nightfall, a preternatural quiet falls with the temperature. Opening the front door at nine in the evening, it's to a kind of silence. It's a city silence, weighty and unnerving, all sudden, unexpected absence. If even a single car grinds slowly past, for once it doesn't seem unreasonable or nosey to wonder why it's out or where it's going.

In the night a few times, I've been wakened by the stillness. Even the gulls are silent. Here, you're often woken by the sounds of gulls. Even when it's nearly morning, in winter darkness, it still feels like night, their cries arching lightly in the air over the silent city. I waken and then as I sleep again, think about the sounds they make which might be of warning or joy, or grief, but which are most probably an unfathomable *Larus* chorus of dialogue and exchange. For me, gulls' voices are a welcome wakening, a kind of wild music, a reminder of where I am in that moment of renewed consciousness: a city on the edge of the sea, at the north-eastern rim of a northern island between the western coasts of Scandinavia and the eastern beginnings of North America, the southern reaches of circumpolar north.

Looking down from a plane window when you're flying towards it from the south, for a long way below you see only rock and grass and fields and suddenly it's there, a tight grey city with sea and water almost surrounding it. It's a city perched on the edge of water, a city of two rivers, blown by every wind named and unnamed, by *ban-gull* and *haugull, blinter* and *flist.* There are days in the wind and rain when it feels as though the whole of it, every edifice and structure, every garden, streetlight and tree will detach and set out determinedly to sea. The grey granite from which Aberdeen is built can look only a semitone lighter or darker than the clouds and sky.

We're part of a thin string of cities, a chain of northern places, poised along this numbered scale, the circles of latitude; the last habitation before the real cold begins, on the fringes of subarctic ice and snow, at the northern reaches of an earth circling in an ellipsis around the sun. We're spinning, tilted on an axis pointing between the North Star and the Southern Cross, tilted only 23.5 degrees, just enough to give us seasons and the words for seasons, for changes in the light and darkness, our hours of day and night. Without the tilt, there would be twelve hours of darkness and twelve of light. The sun would rise and set at the same time every day. There would be differences in temperature because of alterations in distance from the sun but no seasons. Tilting, we incline at certain times of year from or towards the sun. At 57 degrees of latitude, our winter days are short, our summer ones long.

In these northern coastal cities of stone or wood or steel, there's no forgetting the weather or the sea. Sea is in the air, in the mist, in the haar that rolls in fine wet clouds from the coast, up through the harbour, rolling like smoke over the streets and through the lanes. It wreathes round spires, infiltrates the city even on days when inland the sun is warm in a perfect sky. The sea is our orientation, our vanishing point, like the fluid in a spirit level.

It's more than twenty years since I came here to live. I was born a few years after the end of the war in Glasgow, a big city, bleak and blackened then, its porous sandstone walls stained by coal and industry and war. It felt perpetually dark, blighted by eerily quiet, yellow, lung-destroying fogs, and still marked everywhere by the evidence of bombing. On our morning journeys from the suburb where we lived to our school in the middle of the city, we passed a sheared-off tenement wall where, three storeys up, a fireplace hung frozen in some unimaginable moment like an incidental memorial, condemned to face a horizon once red with the fires of bombed and burning Clydeside. The city is clean and changed now, scrubbed bright and golden, entangled with new roads and flyovers. But when I visit, I know the past is still underlying the present as it always is, an indelible imprint of darkness.

On the day in May when I came to look for a house, I had never been to Aberdeen before. From the first, it seemed a city unlike any other, even the other Scottish ones, unlike Glasgow or beautiful, gracious Edinburgh where I lived for years. I remember walking out of the station into a maritime place. It felt Baltic, Hanseatic. I walked out to a harbour, to the sound of gulls, the air blowing with what seemed to me to be the scent of fish and water. The stone was austere and grey, utterly unlike the soft and muted sandstone of those other cities. The broad city-centre streets, the strikingly plain Georgian terraces and early-Victorian cottages were different too, buildings with curved stone gables and oriel windows. I wandered through the network of back lanes behind the city centre with a sense of surprise as if, by some magical mingling of history and infantile solipsism, only my presence had called all this into being.

Geography makes this city easy to overlook, even for Scots, a hundred miles north of Edinburgh and Glasgow, 550 miles from London. (A city more than five hundred miles from London! Is such a thing possible?) It's a place you come to with a purpose. You come to find oil or a house or to sail on a ferry to the Northern Isles.

Shortly before we moved here, I told a friend in London, an Arabic speaker, that we were moving.

'To where?'

'Aberdeen.'

'Ab-ed-din!' he said, delighted. 'This is an Arabic name!'

Aberdeen's small. It's utterly unlike the numerical, geographic giants, the urban behemoths that take a day or more to cross. The very word 'city' itself seems too big for us – cities, surely, should be other than we are. They should be vast and loud. They should spread far beyond the margins of sight. They should be populous, somnambulistic, eternal. Even the words we use of cities seem too weighty for us: *metropolis, megalopolis, conurbation, suburb, exurb.* Aberdeen has none of the sense of frantic wakefulness of other cities, the restless, impatient feeling you have that, whatever the hour, something's happening somewhere.

I get an email from a friend: 'I've just been in Mexico City. It's got a population of 25 million, or something like that ...'

That's 115 times the population of this city. I calculate and make comparisons. How many houses and acres and roads would there have to be? In a moment, I've expanded us. I've spread us a hundredfold, more, over the surface of the earth. I build to accommodate an imaginary populace. The city approaches the Cairngorms forty miles away. It encroaches far out along the banks of both rivers; houses, streets, office blocks fill the valleys of the Dee and the Don. In my imaginary city, you can no longer see from the top windows of my house straight down the city to the horizon and the sea.

'It's amazing', my friend writes, 'how you stop noticing the pollution after twenty-four hours.'

Here, the sea winds blow away at least some of the exhaust fumes. Where they can, they swiftly carry away the accumulations of this city's passionate, unflagging devotion to the car. Here, we live with sound and weather, with wind or rain or the silence of snow. The air can be crystalline but equally it can be pinched and mean until it feels as if we're trying to summon light by looking at a day through dusty lenses. The light here makes the sky seem wide and high. Everything that surrounds us is subtle. Ours is a quiet landscape, a quiet ecology, muted and northern; our bleaching is sea and wind done, sand-worn, impermeable, cold. Brightness blazes round other latitudes, southerly ones, with colour and heat and light but we, like birds with the finest discernments in their many-coned and rodded retinas, habituate our sight to shades and motes of dun and fawn and grey, all fading and understatement. This city is made in every register of grey. Sometimes when sun strikes stone, the mica in the granite casts a million points of dazzle, turning it to a city sprinkled with light.

I found a house that day in May many years ago, and my family – David and our daughters, Bec and Han – lived in it for a while and then we moved away, first to Edinburgh and then to London, and after a few years, we came back again. We found another house, the one that's still home after more than twenty years.

November 28th

At first light the cold seems even fiercer and more implacable than before. I have to smash thick slabs of ice from the doves' and outside birds' water dishes. After I've refilled all the feeders, I search the garden to see who has walked over the snow but the surface is so crisp and hard, there's very little evidence, only the remnants of seed cases and bird food scattered on the ground. The only avian footprints, faint dents and shadows, are wood pigeons', the only birds heavy enough to leave a trace. This morning, I saw a single line of cat prints, which stopped abruptly as if the unfortunate creature had been snatched suddenly, shockingly into the air by hand or beak unknown. Snow must have fallen after it passed by and drifting has obscured the rest of its tracks.

I decide not to use the car at all – I don't use it very much even in fine weather, and the effort of digging it out from the banks of snow piling in the road and gutters is far greater than the effort of walking. I hate driving on snow and ice, all the crunching, the heart-stopping skidding, the sudden slides and sideways drifting. The pavements aren't much safer – they've become slabs of sheet ice. People walk determinedly down the middle of the roads, which are intermittently gritted. Then more snow falls, coating the grit and hardening it into ice. There is a run on spades in the local hardware supermarket. Cold-weather tyres have run out, and snow-chains too. I've never heard of anyone using snow-chains here before. The outdoor shops have no more ice-grips for boots but a friend has brought me some from Edinburgh where clearly they plan for winter better. With these attached, skimming at speed over ice and snow, suddenly I'm Hermes in climbing boots.

Now, I do everything and go everywhere on foot and buy only what I can carry. I work all morning and take time off in the afternoon to walk. Away from the busiest area of the city centre, I'm often the only person in the street. Traffic's sparse. By three thirty or so when it's already dusk, very few cars pass and I'm alone. It makes me think of wars, disasters, prolonged sieges, but it's only snow.

November 29th

Every morning as I sit down at my desk, before doing anything else – including beginning work – I check a website that will inform me if the aurora borealis, the Northern Lights, may be seen later in the night skies above us. The site's called 'Aurora Watch', and is the website of the UK Sub-Auroral Magnetometer Network, SAMNET, the body that, on behalf of us all, keeps a usefully close eye on what's happening in the arcane universe of space weather and on the aurora borealis, the only sign visible from earth of the astonishing turbulences above us. The website has four possible messages: green, which indicates that there's no significant geomagnetic activity and that the aurora won't be seen from anywhere in Britain; yellow, which indicates that there's minor activity and it's unlikely to be seen except in the far north of Scotland; amber, which promises possible visibility in Scotland, the north of England and Northern Ireland, and red, that the aurora may be seen all over the country.

The appearance of the aurora is determined by the interaction of solar wind with the earth's atmosphere. Earth's molten core generates a magnetic field that deflects passing solar winds. Solar flares, a vast release of magnetic energy from the sun, affect earth's atmosphere, causing an acceleration in electrical activity, all of which combine to produce the aurora borealis. It seems, for reasons no one can yet explain, that the aurora appears more frequently at equinoxes, particularly the vernal.

Not long after we came here, we used to see the aurora. It wasn't a frequent occurrence, but was more frequent than it is now. Before I began my quotidian checking (in those days that now seem long ago but aren't really, when websites didn't exist) I relied on watching the sky, on friends phoning to say that the aurora could be seen from whatever part of the town or country they were in. If it was near enough, we'd get into the car and drive out to experience this most wondrous of sights. We'd see it sometimes from the back garden too, this startling shimmering of light, pink and yellow, green and gold, falling over us like a strange ethereal curtain flung down from the high atmosphere, a soft canopy of moving luminescence dropping from the sky.

It's a long time since I last saw it, even though the aurora is one of the things for which this city is known. It may be known for other things too but the ones for which it is famous are the name it has given to a breed of fine cattle, for oil rigs and platforms and pipelines and for dazzling lights in the sky.

For years, I watched for it. Sometimes on clear nights, the light through the window blind in the bedroom would seem to shimmer and move in waves, and I'd pull the blind aside to check, but the sky was always dark and lightless. Over time, it became more common for public buildings and offices round about us to direct powerful beams of white and blue and even green light over their facades, for security or vainglory, lights that, all over the world, in every city, prevent us from seeing the sky. I'd look from the back windows onto the garden, still a small area of tree-protected darkness, down the line of unlit gardens and office car parks, or I'd stand in the middle of the grass, looking up, but while I might see a breathtaking spray of stars, or a lambent, brilliant moon, I didn't see the aurora.

Then, after a long time of not seeing it, I began to wonder, how common were sightings here? How often was it seen in the past?

Named by Galileo after his observation in 1619, '*aurora boreale*', 'northern dawn' and thought by him, wrongly, to be the sun's reflection on the earth's atmosphere, the foundation of many of the creation myths of northern peoples, the aurora has been interpreted and seen as shoals of herring in the sky, as swans in ice, flights of migrating geese, fire sparking from the coat of a fox, and in Inuit legend, as the spirits of the dead playing with the skull of a walrus – all this to explain sheets of light moving in the sky.

It makes me wonder how long it takes for a natural phenomenon to be incorporated into the cultural history of a place, referred to in song, adopted in the naming of pubs, bands, petrol stations. The perception that there were simply fewer sightings wasn't, as I had suspected, just an incomer's perception – people who were brought up in Aberdeen commented on it too and we wondered if, for all of us, it was a trick of memory, a golden-age imagining of natural and altogether past perfection, or perhaps of childhood, those illusory moments from which we're putting distance, all too fast.

The question is on my mind at a gathering of friends when we begin to discuss the lack of sightings of the aurora. We all refer back to the fact that there's even a song about it – 'The Northern Lights Of Old Aberdeen' – as maudlin and sentimental a ditty as anyone ever heard sung by drunk men anywhere. We decide that no one would make up a song on the basis of nothing at all, or not a song that sounded as if it was about something quite so specific. Would they? We begin to discuss the words, wondering if there might have been a misunderstanding, if they might refer to the terrestrial lights seen from a distance by the seafarer approaching his home from the sea, and not to the aurora.

Although sober, we each take turns to sing the bits we can remember of this dismal, unmelodic work. The chorus we recall as being something to do with a traveller longing to see the lights of Aberdeen although nobody can remember the exact words. One person believes that in the verse, there's mention of the Northern Lights looking like heavenly dancers in the sky and so, while avoiding discussion of the artistic or musical merits of the work, we all agree, city and country dweller alike, that there can be no doubt about the fact that the aurora was an established fact of Aberdeen life, and that none of us has seen any celestial dancing, or not of that particular sort, for years.

But solar weather operates in cycles, and it seems that one is ending. It would just about coincide with the paucity of sightings. The words that describe the inception of the aurora – *auroral ribbon, magnetospheric sunstorms, solar wind, sudden storm commencement* – are almost enough in themselves. They're inspiring, exciting, these words of passion flung up towards a possibly deceptively peaceful sky.

Today, as usual, I'm told that here at least we're not going to see it. 'No significant activity', the message says. Though it seems worth the vigilance. There are specific features of life in northern latitudes: the weather, naturally, a kind of ley line of proximity, our closeness to Scandinavia, or our inclusion in the magnetic sphere of the aurora borealis, that create a sense of otherness from the rest of the country. This is a place where, like no other, one pays attention to the smaller changes of light and season, to solstice and equinox, the lengthening and the shortening of days.

All morning as I work, I hear Ziki the crow calling as he always does

27

when there's snow. He's looking from the window of the room where he lives on to the snow-lit garden, listening to music and cawing loudly. After years of there being many birds and creatures in the house, the numbers are reduced. Now, in addition to Ziki, there's Bec's cockatiel, Bardie, the first bird we got, and there's the rook, Chicken. Both have lived here for almost as long as we have. Ziki the crow is only four, the only one who, in bird terms, isn't old.

I'm distracted, sitting at my desk, turning constantly to watch the altering sky. After a clear blue dawn, it deepens to azure in the east and over the next hours, fades to chalky, icy white and then it snows again. This time, it's not the big, slow flakes that fall grey on white against the sky. These are small flakes, grains of snow, quick falling, a fast unburdening. I feel as if I'm just beginning to learn the subtleties of snow. I become aware of a robin who has come to sit on the other side of the glass. He's on the sagging canopy of clematis that has been unmoored by the weight of snow. I can see a dust of whiteness on his breast feathers, the swiftness of his breathing.

The morning darkens down and down and still more snow falls. Outside the study window, a grey city is turned by the light into a city from an Edwardian silver print of roofs and walls and branches; a heavy, tinted sky. I take a photo on my phone from the study window to send to a friend in England, an Aberdonian-in-exile.

'Just to show you what you're missing!' I write, a fatuous enough message. A few seconds later, a brilliant point of light flashes into my eye from outside. I'm startled for a moment. In this inexplicable parallel moment, it seems as if someone's taking a photograph of me, someone unseen, perhaps at the bottom of the garden. What is the flash of light? Could it be lightning? I know it is only when, a few seconds later, it begins to thunder, a sound as large and dark as the sky. Snow and thunder. How?

In fact, it's a rare, named natural phenomenon, *thundersnow*. (It sounds like the title of a forgotten classic of northern European literature.) It happens, I discover, when moist air, cold enough to produce snow but warmer than the layers of air above in the troposphere – the lower portion of earth's atmosphere, the bit where weather happens – rises strongly, forming snow and ice inside unstable clouds, triggering thunder and lightning, when snow falls instead of rain, delighting the hearts of meteorologists everywhere. And my heart too. It's magnificent, exhilarating! Thundersnow! (Rare though it is, it happens again not long after when snow falls in January over Alabama.)

December 3rd

This morning, I'm driven to a quiet place in the Aberdeenshire countryside twenty miles away, to take part in an item for a radio programme about rooks in snow. They've been seen picking at something in the middle of roads and Mark Stephen from the BBC and I go to find out if they are, and at what, and why. The wide, usually busy dual carriageway is reduced to a thin, one-lane passage between low walls of snow. Outside the city, it's an ice world of frozen fields and trees and sky. Icicles like eerie, glittering fortifications hang from city roofs but in the villages they've solidified to long and glistening spears of frozen water almost reaching the ground. Beyond the windows of the car is a make-believe world of mythic winter, where palisades of ice seem to be defending fairy-tale cottages, forming bars against the doors of witches' houses. (Idly, I wonder if an icicle would make the perfect murder weapon. Other people must have thought of it, in literature anyway, but I don't know if anyone has tried it. It reminds me of our magpie stealing and hiding ice cubes – the effects are the same whether it's booty or evidence – ice melts.)

Fields have become smoothed-out plains of glistening, crusted ice, stilled surfaces glancing a fierce and blinding dazzle in the cold, brilliant sun. Hoar frost coats forests, whole trees, single branches, glinting, sparking blue and white and fire from the light and air. Every soft fall of snow from branches explodes to mist in a rainbow of fine-hazed spindrift. We stand on a farm road in the ringing silence and talk of the effects of road grit and road salt on birds, and of rook behaviour and, tangentially, about a type of motorised American bird-feeder that springs into action when a squirrel tries to feed, flinging the unfortunate beast off the feeder or tipping up the perch. Unhelpfully, there are very few rooks to be seen. It's only on the drive back that we see why – a manure heap we pass is teeming with rooks seeking warmth and any insects there might be in this suddenly changed and inhospitable land.

December 7th

Since being outside the city, albeit briefly, I've felt even more sur-rounded and enclosed. This morning, I sit at my desk and take from the shelf my old, worn copy of *Tao – The Watercourse Way*, a book on Chinese philosophy by the great Zen and Taoism scholar Alan Watts. The quietude and stillness seem to encourage meditative thought, or else it's just one of the moments of guilt I experience from time to time at having forgotten quite so much of what I learned long ago.

> Certain Chinese philosophers writing in, perhaps, the fifth and fourth centuries explained ideas and a way of life that have come to be known as *Taoism* – the way of man's cooperation with the course or trend of the natural world, whose principles we discover in the flow patterns of water, gas and fire which are subsequently memorialised or sculptured in those of stone or wood, and, later, in many forms of human art.

Watts explains what he believed to be the importance of these ideas for our time, a time when 'we are realising that our efforts to rule nature by technical force ... may have disastrous results'. (The book was incomplete at the author's death in 1973, and was completed by his friend Al Chung-Liang Huang and published in 1975.)

This morning, I reacquaint myself with the ideas of yin and yang, the forces that are part of all life on this earth, the principles of polar-ity and balance, the existence of both in all things, ideas so much at variance with Western ideas of opposition and opposites, conflict and vanquishing. I reread some of the writings of Lao Tzu, Chuang Tzu and Lieh Tzu and remind myself of the central Taoist principle of '*wu wei*', translated and explained variously as 'masterly inactivity', 'not forcing', 'knowing when not to act'. It seems like a good idea. It is illustrated by Alan Watts with a metaphor that feels strikingly appropriate, one of branches of pine and willow under snow, their differing qualities of breaking under weight, or bending, springing back.

Reading about the Tao, 'the Way', I'm far from sure what it is, of what significance there might be in snow, floods, rain. There's no way of knowing if one year's cataclysms of climate might be related, or even portentous but the words I read speak of wholeness and a different way of living, of perceiving the world and our actions in it. I don't know if even thinking about it is escape, or hope where there is none, or if trying to look equably at the world is doing the only thing one can.

Alan Watts's advice to his readers in preparing for an understanding of the text to come, is: 'Take it that you are not going anywhere but here, and that there never was, is, or will be any time but now,' and he suggests putting aside for the moment all opinion, all knowledge except in the interpretation and acknowledgement of sensation, to appreciate what is without giving it a name, to listen and see and breathe in a state without word or thought.

'Stop, look and listen ... and stay there awhile before you go on reading.'

December 11th

Often when you're walking through the city, you can hear unseen water. It gurgles unexpectedly under one crossroads, gushes strangely below my feet at another. As you walk on, the source of it appears – the Denburn, flowing through deep-constructed channels beside a row of offices before it disappears again under the road, through culverts and on towards the harbour where it enters the sea. It's heard best after heavy rain, even from behind the fortifications of the large granite houses with their long gardens in the steep declivity of Rubislaw Den where it runs through the gorge formed by glacial meltwater, a valley that bisects the centre of the city. You hear the burn in spate on a cool evening in summer after rain when light shimmers on the summer leaves and everything's heavy with dampness, but now there's only a still and freezing silence.

Today, I walk for a while along the banks of the Dee near Duthie Park. The margins of the river have frozen. Broken, milky patches of ice drift in the black water. Gulls stand massed on the shingle islets. Their feathers lift easily from their head and wings, white-fringed in the freezing winds. Waxwings flit and scatter, searching for berries on the cotoneasters and holly in the snow-covered gardens.

On the way home in the afternoon, I find a fledgling in the lane. In spite of my decision to abjure bird-rescue for ever, I bring him home. There's nothing else I can do. He's a small pigeon with a damaged wing. I examine the wing, which doesn't look very bad. If he survives, I'll wait until he's ready and let him go.

December 12th

Every day the snow seems still more churned and frozen. A new pavement landscape appears: a fresh set of ice ditches, ridges and mounds form and freeze. Everyone has to pay close attention to the elemental business of putting one foot in front of the other. Walking for a long time in snow seems to affect different muscles and makes me aware of small alterations and tensions in the legs and hips, a bit like the effects of walking for a long time on sand.

December 20th

Tomorrow morning, there's to be a total eclipse of the moon.

MIDWINTER

MIDWINTER

December 21st

Midwinter, early morning and a clear and perfect sky. By the time I look out, the penumbral phase of the eclipse is already over and the partial's ending, the moon moving towards totality. From some places it'll be seen as a deep and burnished red, but from here it looks pale, a delicate pink-silvered disc in a cold-weather sky of clear and glowing indigo. If I hadn't heard about it on the radio, I wouldn't have known. I might have looked out at the quiet winter moon and not seen it slip into the umbra of the earth or known the facts of its happening, a startling calculation that seems to give the event particular significance. This is the first lunar eclipse to fall on the day of the Winter Solstice for 372 years, only the second since the moment Western civilisation began its measuring of years, since the birth of Christ, the beginning of the Common Era. I'm lucky – I haven't had to do any of the things many people across the northern hemisphere have had to do to see it, to drive out into the countryside beyond the lights of cities to watch and wait, to lie on my back, telescope pointed expectantly towards the momentous firmament, urging away the clouds. I've only had to climb the stairs to a window on the top floor of the house and take up position on the sill to gaze towards the south-west. Below me, the snowbound city a few days before Christmas. Christmas-tree lights sparkle from a few windows. Today will be another freezing day of brilliant sunlight, the shortest of the year and dark by half past three. Although it's the day of the Winter Solstice, the precise moment when the earth's axis is tilted at its farthest point from the sun doesn't occur until 22 minutes before midnight. It's the day of solstice that's celebrated, the idea of it, the symbolic progress from maximum darkness to the light ahead. This year began in snow and now, it looks as if it'll end in snow.

Through my binoculars, the moon's transformed from an insubstantial, glowing circlet to the solid sphere of rock that it is, tinted rose by dust-filtered sunlight high in the sky, as north as it can ever be, somewhere above Orion the Hunter. I think of every philosophical or metaphoric quality that has been bestowed on this distant object through the history of human longing or describing, and of the band of other watchers all across the seeing side of earth; their eyes, like mine, fixed on this glowing orb 200,000 or so miles away, moving in its entirety into the shadow of the planet where the life of everything we know exists. Sitting in my pyjamas on the sill, I watch the moon move west across the sky towards the moment of greatest totality.

('Carry enough equipment with you, including sleeping bags, blankets, extra batteries, heavy-duty gloves. Don't let your hands freeze to your equipment,' one American moon-observer's website suggests; 'eclipses take hours. Keep warm.')

From the hall two floors below, Chicken the rook calls. I call back. I tell her that I'll be down in a minute although I know that, given the propensities of eclipses, I probably won't be.

As the moon moves across the lenses of my binoculars, the morning light increases. Two of the jackdaws who live in nearby trees land lightly on the metal arrow of the weathervane on the finial of the roof across the road. They take up position, one east, one west, and perform the ritual preening and calling of morning. The occasional gull passes in slow grace over the roofs and houses.

The eclipse seems at first to be the main event, and is, in the way cosmic forces are, determining by their existence the nature and even the fact of our own, in the movement of sun and moon and stars, commanding tides and gravity, light and warmth, all vital to our lives but largely unnoticed except by those who, for millennia, have watched and charted, their minds enviably filled like those of birds with the trajectories of planets and the paths of stars. But here on earth, the main event is the day. It's still quiet in the streets below but soon, it'll be as any city on any morning. Today will be about negotiating snow again: sledges and cars and children, people dressed as if they've effortlessly become citizens of a different kingdom, one of snow and ice, and have adapted to a new climate, one which might not be a new climate at all but the remnants of an old one, or simply the inscrutable, fathomless way climates have always been.

As it moves west across the sky, the moon doesn't seem to be travelling at 2,300 mph – a kilometre a second, slowed by distance to a deliberate serenity. Earth's shadow falls across it, the shadow of the planet on which we live, the shadow of everything we were, are and do.

Downstairs, Chicken's becoming impatient. Even in old age, she's hyper-aware of time, and recognises eight o'clock in the morning more accurately than any sophisticated timepiece. As the moon drifts across the broad, clear sky, I call to her again to tell her that I won't be long. The eclipse is ending, moving from totality back through the stages of its reversal. I watch its progress as far as I can, until I have to stand on tiptoe on the broad sill to see the last of the moon as it slides behind the winter branches of the beech and sycamore and pine in Rubislaw Den and melts invisibly into the morning light.

''Tis the yeares midnight', John Donne wrote in 'A Nocturnall Upon St Lucies Day, Being the Shortest Day'.

'... The Sunne is spent ... The worlds whole sap is sunke:'

It is the year's midnight. The world's whole sap does seem both sunk and frozen when, shortly before the day ends, the moment of solstice occurs and I'm asleep.

December 22nd

The days move on from the eclipse, still in snow. The small pigeon flies off early one morning without my having to decide to release him. He simply appears to have decided it's time to go.

The peace the snow has brought has been disturbed over the past few days, perhaps by the approach of Christmas and its attendant anxieties, or because with snow-chains and snow-tyres, people have become more confident about driving in these conditions. As I'm crossing the road near the theatre, a car's stuck on the slope on the corner, its wheels churning ineffectually. The man in the car behind sounds his horn. The driver of the first car gets out and shouts in understandable frustration, 'I'm stuck! What do you want me to do?' and behind her, people open car windows or get out to see what's happening. A line of anxiety sparks like a live cable down the queue of traffic. Some passers-by rush to push the car. Everywhere, cars are being manhandled out of drifts, their drivers turning anxiously as they drive off at last, feeling guilty that they didn't have the chance to thank their rescuers.

December 23rd

The family arrive home, Bec and her daughter Leah from Edinburgh, Han and Ian from London.

December 24th

Han, Ian and I walk into town. As we walk, we notice that suddenly the snow's loosening, the ice is softening, glistening wet. Our feet make indentations in the slush as we walk. The thaw has begun. It feels no warmer.

December 25th

Christmas day, still white but gently dripping.

January 3rd

Everyone's gone home, and the house is quiet again. It's not unwelcome after years of activity and voices when it was occupied not only by people, but by animals and birds, when there were shrieks and calls from every corner. Chicken is still vocal, and Ziki, and even Bec's elderly cockatiel can still summon himself to a bout of lusty shrieking although less frequently now than in the days of his long-ago youth. From the room where I work, in summer when the door's open, I can hear the mutterings of my elderly doves in the doo'cot, but other voices have gone. These days, no magpie passes me, shouting, on the stairs. The rooms are silent and empty, all the places where once birds and creatures lived. No rabbit climbs through the kitchen window to glissade down my printer as I work. The only rats are wild ones who, from time to time, take up residence in the space under the house, or ones brought for visits by Bec, who still keeps them as pets. (Although they didn't contribute to the noise, they contributed in immeasurable ways to the life of the household.)

I miss them all, yet I don't – they seem now part of another life. I see the passage of years as a folding-up of time. Synaesthetically too, I see it in sound, a large, quiet house filling with noise, bird and human, with song of one sort and another, incantatory, celebratory, complaining, with periodic changes in chord, night songs and morning songs, warning calls and greeting calls from an assembled company, many-specied, assorted in character and colour, habit and tone, as diverse and serendipitous an assemblage of orphans and foundlings as might have been gathered together in any Victorian workhouse. I see the rise to a loud and often disharmonious crescendo, the fall into diminuendo, or rather, the small diminuendos of individual lives and deaths.

January 4th

And after the lunar eclipse, that terrific, super-scale meeting of history and planets, only a fortnight later there's a partial solar eclipse. (They occur like this, in pairs, two weeks before or after the new moon.) Busy skies! Busy planets! This time, I read that the sun is to rise at 8.46 a.m., which it does, in another clear and perfect winter sky. I can just see its low and stealthy glow over the shadowy garden, the reflection of light on clouds, the quick glint of a window far down the valley of lanes that lead in breaking lines across the city towards the sea. The wings of a gull dazzle with sunlight for a moment as it passes overhead but I can't see the sun. I have my eye-protecting X-ray film ready, kept since the total eclipse a few years back in anticipation of all those yet to come, but the sun's too low in the sky for me to need it. I think that I can discern a small difference in the light, some sombre, subtle gradation of tone, but probably it's only because I know the eclipse is happening. By the time the sun's high enough to be visible, it has already passed out of eclipse. It's as high as it can be in a winter sky, and riding free.

January 5th

It's still only the first week of a new year. How quickly all the celebrations, all the effort of Christmas and New Year, disappear into the slipstream of everyday life.

Helen D and I drive out a little way to walk along the beach near the city with Molly, her lurcher. Today, we're glad to be out, away from stoves and fires and the confining cold of the past weeks although my experience can't compare with Helen's – she lives inside the boundary of the Cairngorms National Park, forty miles away, high up towards the mountains at the end of a rough track. When she came to visit one day last April, she had to ski down the track to her car.

We trudge down through the steep paths in the sand and high dunes towards the beach. In the icy wind, the sand's too damp to blow, tamped down, saturated by the weight of the months of snow and snow-melt. When we get beyond the dunes, the long beach is empty. Afternoon on a day of crystalline chill, light glowing from under pale, translucent waves. The North Sea isn't warmed as the seas of the north-west coast are, by the Gulf Stream. It's cold beyond breath or mind although, unimaginably, people do swim here on New Year's Day. This is sea that can kill you quickly, winter or summer. Beyond us, strung along the coast to the north, are all the fishing towns, Peterhead and Fraserburgh, Buckie, Portsoy and Cullen, places that know well the closeness of sea and death.

It is still bright but the afternoon'll go fast. The lights of the city will come on soon behind us, the lights of rigs too, far out at sea. We walk north. A few herring gulls seem to move with us in their drifty, serene, unhurried way over the water. It's a clean beach, with little of the usual detritus the sea washes up. There are no plastic ropes, no fertiliser sacks, no plastic bottles.

Molly runs ahead, looking absurdly like Helen's sculpture of her. Helen works in metal, and seems able to create life and the movement of joint and muscle from fire and bolts and scrap. I expect Molly to transform in a moment into stillness, to be not her beautiful, lithe and running self but Helen's making of her, all watching, tensile power, near to flight.

This beach isn't remote. It's near the city, where generations of Aberdeen teenagers have come for summer picnics, where city people walk their dogs. When you walk here alone, it's easy to feel detached and far away, between earth and sky. Here, remoteness seems like wildness, something in the mind. It's a beach that makes me think of the American writer Henry Beston, who in 1926 spent a year living in a small house he had constructed on the dunes at the Great Beach of Cape Cod. From his house, the Fo'castle, he observed the life of sky and sea, of fish and birds and stars, writing of it lyrically in his book *The Outermost House*. The idea seems fleetingly attractive but then I think of the length and cold of winter.

Today, the sea is like a Flemish maritime painting, a canvas of radiant, luminous afternoon light, Simon de Vlieger on a calm day, but it can turn bleak quickly on this coast of dune and sand and cliff, wild and cold and blown sea where everything feels limitless, beyond time, this coast created by an eternity of sand and wind.

I think of words of tide and water: *marsh, littoral, estuarine, lacustrine*; of the marks and effects of time. This dune system is ancient, formed by melting Holocene ice, high dunes and sand sheets, structures layering, rising, moving, shifting, mobile and alive. Four thousand years of blowing sand. People have written of the unique features and importance of this place: 'marram-adorned, extending in an unbroken cordon ... shore parallel dune ridges, dynamic sand sheet, migrating north; internationally important', it says in a report on the geomorphology of the area. Another says: 'This site's assemblage of natural windblown landforms – a sand sheet and inundated dunes – is of an extent unparalleled in Great Britain.' It's so valuable and rare that it has been designated a 'Site of Special Scientific Interest'.

The only human sound is of a single engine somewhere to our left. Second World War tank traps still lie half buried in the sand on this sea-edge facing out towards Norway. It's clear to the horizon in the cold but in summer, with any warmth, the haar rises from the sea and loses the coast and city in a pale tide of mist.

Ahead of us are miles of sand and dune until the flat coast rises to steep and terrifying sea cliffs. This is a coast of birds, of residents and migrants, of passing feeders, those stopping on the way to somewhere else, flying north to Forvie or the Ythan Estuary or the Loch of Strathbeg. It's a place of skylark and meadow pipit, lapwing and plover. There are linnets and reed buntings and golden plovers, snow buntings and terns, the waders too: the dunlin, merganser, eider, widgeon, oystercatcher, the sanderling, goldeneye, redshank, the swan, the mallard, the greylag and pink-footed geese, the cormorant. It's a place of willows and orchid, iris and fern, of mosses and lichen, of butterflies, badgers, otters, hares, all of them almost within the ambit of city lights.

As we're walking, from nowhere, from the air or sand, four small grey and white birds appear in blue shadow, skimming the shallows, heading somewhere at miraculous speed. They move so fast their legs are invisible, part of the sea itself, part of the liminal afternoon light; sanderlings, *Calidris alba*, wading birds who breed in the High Arctic and winter here. (This, then, is the equivalent of their summer holiday.) They're known for rushing into the waves to feed on the minute creatures brought in by the tide but today they run on past us, full of avian purpose, resolving into fine black dots in front of us. Sanderlings have no hallux (back toe), although how you might tell from watching them running by like this, I have no idea.

Walking on this January afternoon, I see us as we've walked here on other days, in other seasons, with other people, with banners and chants and singing, part of a ragged protest army straggling through the dunes, someone playing the pipes, someone singing, but today, we hear only the single engine of the digger we can see once we've climbed up through the steep channels in the dunes, past the wood and wire fencing erected to keep us out. A small, single yellow digger, a child's toy in the vastness of the dunes, working, scooping a single load of sand into its bucket, heading towards the accumulating pyramids as it digs up the dunes to turn them into a 'world class' golf course.

The afternoon dips from luminescence towards the clear deep blue of dusk. Walking back, the wind hits us but today the sand's too damp to blow. When it's dry, it sweeps over you, whips past you like veils of quick-blown chiffon, a storm and tide of sand.

Soon, the dunes will be gone, the Site of Special Scientific Interest, the breeding birds, the wintering waders.

Helen and I talk about the moment months before when, as one of the protests was ending at a farm nearby, all of us were standing together, gathered to try to save this place, and two great whooper swans circled twice round us as we stood. We talk too about the village of Forvie a few miles north and of the wild, unrelenting wind that, over the course of nine days 600 years ago, buried its community and its church in sand. It's only what people say will happen again one day.

As we're driving back towards the main road, a flock of curlew rises from the low winter fields, their dappled plumage bright in the last flare of day.

January 18th

With the ending of the first phase of winter seems to have come the ebbing away of our need to exchange snow tales, the ones we told like redundant polar explorers. Mine, being city stories, were always bound to be lesser, anticlimactic, frankly small: 'I made it to the library!' (Sturdy northerners as we may think ourselves, sniggering at news reports of London shutting down as a result of what we regard as minor falls of snow, we overlook the gridlock that occurs immediately in this small city every time more than a few flakes fall.) The weather here is a big topic; sometimes, the only topic. Even mentioning it marks me off from the vast majority of humanity that does not live in the United Kingdom, or doesn't think weather worth talking about, the portion that may still regard weather with equanimity. (Once, on an astonishingly, blastingly cold February day in Budapest, I discovered the follies and humiliations of attempting to discuss the weather when anywhere other than Britain. The mild and affable comment I made about the cold to the person selling tickets in a museum was met with silence and a gaze as glacial as the temperature outside. Of course it's cold, the look said, it's February. *Idióta*.)

We talk about it here because it changes, it flits and blows, defies forecast and forecasters, ruins the careers of those who have stood in front of an audience of millions, asserting with misplaced confidence exactly what the next day's weather is to be. Here, there may be several seasons in a day, only some of them recognisable, only some of them conforming to the words we have available to us to describe them.

In Scotland we have religion to convince us that however bad the weather, we deserve it. Every adverse weather condition is a punishment, and moreover, a deserved one. Every fine day is a boon, an entirely undeserved one for which, inevitably, we will pay a preordained price. (There seems to be a latitude of religion, too – we may all be penitents above 55 degrees north.)

We discuss it. We attribute a lousy summer to our own hubris, to the optimistic purchase of a summer dress, a pair of sandals. (Last summer's dismal weather, without any doubt, is a direct result of my having bought new garden chairs.) Pleasure is diluted by presentiment. Alastair Reid wrote with glorious accuracy in his fine poem, 'Scotland':

It was a day peculiar to this piece of the planet,
when larks rose on long thin strings of singing
and the air shifted with the shimmer of actual angels.
Greenness entered the body. The grasses
shivered with presences, and sunlight
stayed like a halo on hair and heather and hills.
Walking into town, I saw, in a radiant raincoat,
The woman from the fish-shop. 'What a day it is!'
cried I, like a sunstruck madman.
And what did she have to say for it?
Her brow grew bleak, her ancestors raged in their graves
And she spoke with ancient misery:
'We'll pay for it, we'll pay for it, we'll pay for it!'

We may well pay but despite everything, we adapt. Small changes change us. We're promiscuous in our responses to weather. A few consecutive days of heat (or comparative heat – the highest recorded temperature here that I can find, and I assume it to be reliable, was 27 degrees Celsius) and suddenly, we're expansive. We become almost southern and passionate. We sit in outdoor cafés, or rather ones where tables, chairs and sunshades are hastily rushed out onto any available sliver of pavement at the first glimpse of sun, and either hastily rushed in again when the ephemeral moment passes, or remain outside in sad, unlikely hope, gathering snow, rusting or rotting under waterfalls of rain. Even on chillier days we dress optimistically, or some of us do. On days deemed hot, the trusting or dedicated expose their pale, pale bodies, their polar-white, luminous Scottish skin to uncertain air, to brief, unmediated blasts of raw and dangerous sunlight. After a weekend of heat, the city renders up its puce and blistering. After all, it may be the last sun we ever see, or the last for a long time. *Carpe diem!* Perhaps burning is suitably symbolic, both penance and reward, an anticipation of hell's fire without the trouble of ever having to get there.

Usually in winter, particularly for evening, there's no cold-weather dressing as there is in continental northern European cities, where people appear to notice and even to experience the cold. There, they dress for the cold, fortify themselves against it. Here, we dress for summer in any temperature above freezing, all sleeveless or coatless or vertiginous heels, doggedly, as if the price of hypothermia and the merest touch of frostbite is the required restitution for a good night out. But then, after a few days of severe snow and intense cold, we're booted and trudging, as fickle as the wind and rain. As the odd stasis in the days sets in, along with it, as quickly as we became passionate we become phlegmatic, accepting. Unusual social exchange begins to take place in the icy streets. On a freezing morning, a woman making her way with difficulty through deep, fresh snow smiles at me. I smile back, silently celebrating this new, cautious bond, the bond of very mild hardship, happening in a city not known universally for such warm-hearted interaction. Stopping briefly, the woman says, 'I've so many clothes on I can hardly move. Don't you find that, in this weather, all pride goes?' I say that I do indeed although, in fact, I've never thought about it at all.

January 29th

Low light on a winter afternoon. I go to a gallery to see an exhibition of photographs. The town's quiet. Two gulls cross the empty, cobbled square of the Castlegate. They look as if they're in conversation, like two nuns, their feathers like wimples blowing white in the wind, treading piously on their big, light feet.

In Marischal Street, the town's pigeons are beginning their roost, edging the Georgian roofs with a frill of tail and feather. It is a steep street, sloping down to the harbour and seems to have ships docking at its foot, all masts, reflections of water and moving lights. It's a street of eighteenth-century town houses with an air of the past that seems alive as I walk this ancient route from the Castlegate to the harbour. (It's easy to imagine, to be instantly in another time in Marischal Street on a dark, January afternoon.) I walk across a rare architectural innovation, an eighteenth-century flyover, the bridge built to carry the street across Virginia Street below. It was modernised in the 1980s, Virginia Street widened to carry the harbour traffic that now whisks by on four fast lanes. I look over to the traffic, the warehouses, their names the only constant in the long progress of time: *Aberdeen Shore Porters Society, established 1498*, Baltic Place, Shiprow. (Here, severe cold is always described as 'Baltic'. 'Baltic today,' we'll say, perhaps with the long-held memory of early trade and our Hanseatic past. The vehicles of Aberdeen Shore Porters, now a transport company, bear what seems like the enigmatic comment: 'Aberdeen, 1 Baltic Place', but is, in fact, their address.) I carry on walking down to the quays, Regent Quay and Trinity Quay, to the streets of ship's chandlers and tattoo parlours, maritime insurance companies and fishermen's rests. Gulls flicker white in the lights of the harbour.

From the short brilliance of Solstice Day, the day has expanded to eight hours and fourteen minutes on the steady minute-by-minute progression towards summer and its eighteen hours of daylight. It's twilight. There seems to be so much of everything here. There are different kinds of wind, different kinds of snow, different kinds of twilight. There is, I discover, civil twilight when the sun is 6 degrees below the horizon, when you need light to read outside; nautical twilight, which lasts the entire night, when the sun is 12 degrees below the horizon; astronomical twilight, when the sun is 18 degrees below the horizon. I don't know which this is. Even now the harbour's busy, lit and bustling, vessels getting ready to sail out to rigs at sea, the huge Northern Isles ferries preparing to head out north to Orkney and Shetland.

On the way home, I pass one of the busiest crossroads in the city, where people used to stand to wait for the daily preparation of the starlings for their evening roost and the sky would be scattered from every direction with dark ingatherings of tiny, darting urban commuters making their way across the city to their meeting point above the bridge where they'd huddle together for warmth and the noisy, structured companionship of avian societies. Flying in their busy, purposeful parties, they'd join in a vast and gorgeous display, a sweep and flight that transformed the sky, the city, the lives of the observers with the inexplicable mystery of their precision and grace, with the sight of this moving grey wave, this cloud of birds closing and spreading, dipping and soaring, thousands of *Sturnus vulgaris*, those quick and lively birds, those beautiful creatures of oiled, shimmering gold and green.

I'd stand, breath held, sharing in the birds' joyous flight, in their song, their exuberance and energy until the moment when, as one, they'd turn, their undulating tissue of bird-cloud folding, narrowing into a streamer of molten darkness to disappear under the parapets of Union Bridge.

As I walk back, it's through the noise of traffic alone. The dusk air's emptier now. The song of winter evenings has been diminished. A few years ago, someone at the council decided that starlings were damaging the paintwork of the bridge, that their excrement was blocking land drains, that they were a nuisance and had to be removed. 'Control' methods were implemented and netting installed under the bridge to prevent the starlings from settling.

It's almost always hygiene that leads to these extirpations. Birds excrete, and their excreta collects. It can damage stonework and metal but can also be removed and cleared away. The threat of zoonotic diseases is often used too as a reason to remove the habitats of gulls, starlings and pigeons but it is a threat that is vastly exaggerated. (Diseases! Spores! Bacteria! All carried by every darn thing that moves or breathes.) The incidence of histoplasmosis, which people can catch from starling faeces, is extremely rare. It is a disease caught mainly by people working in areas where droppings have accumulated, and is easily prevented by the wearing of appropriate facemasks. (You might get it more readily if you go caving, or live in the Ohio River Valley.) Zoonotic diseases do occur and many can be dangerous but, in this case, protection of the populace from the unmeasurably small chance of contracting the disease doesn't seem to justify what's been done.

I look up as I walk. A few starlings fly above the bridge in groups of ten or twelve. They look lost, disembodied, as if they've been broken off something larger, something whole.

Even if I hadn't known one personally, I'd have a special fondness for starlings. They were the first birds I knew, the first birds to make their vibrant, noisy existence known to me in my childhood, ubiquitous in the city, in crowd and chorus, their singing lighting the darkness of the Glasgow dusk. The starling I knew personally was Max, a bird found and reared by an American family who couldn't take him home with them when they went back. (The United States doesn't know what to do with the many millions of introduced starlings they already have.) We took him. I think of the years he lived with us, of his excitements and his irritations, his swearing (of the serious but not entirely discernible sort), his sotto voce mutterings, the instant connection he brought to me with a long-ago past. I think of the nature of his character, the exquisite sweetness of his evening solos as well as the extraordinary beauty of the bird, the gilded feathers, the neatness of wing as he flew around the house. After I got to know him, I'd look anew each evening at the cloud of swirling starlings, understanding that each one of them was as Max was. Knowing increased my amazement at their individuality, at the magical coordination of their movement, the singular, transcendent beauty of this turning, sweeping cloud of birds. I used to wonder if they looked down from their elevated high-flying towards those of us watching from the pavement, and see only undifferentiated members of another species.

When I get home, I look out the poem 'The Starlings in George Square' in which the late Edwin Morgan, Makar of Scotland and a fellow Glaswegian, writes of the council that complains of the noise and the mess starlings make, and employs the 'bird-men' to remove them; of the amusement of people who watch the birds trying to land on the plastic rollers they've installed:

'The Lord Provost sings in her marble hacienda … Sir Walter's vexed that his column's deserted.'

Starlings, despite congregations of large numbers, are critically endangered in Britain. Their numbers have declined by over 65 per cent in the past 30 years (as with other birds whose numbers have similarly declined, no one quite knows why) and now their habitat, the foundation for one of the marvels of the natural world, has been destroyed, probably for ever, and I wonder how future generations will learn about the value of the life around them, of birds such as starlings. How will they know what they've lost or are losing? Who will teach our children what starlings were? When Edwin Morgan asks if we really deserved the starlings, it is a question I ask too, knowing that his starlings were my starlings, and all his sadness and anger mine. He writes of the joy of the child whose father encourages him to look at starlings, a sight

that pierces the boy like a story,
a story more than a song. He will never forget that evening,
the silhouette of the roofs,
the starlings by the lamps.

February 2nd

Snow and ice don't let go easily. The temperatures have risen more than twenty degrees from their lowest point and so, even though it's still very cold by comparison with the past few weeks, it feels almost mild. Minor ice floes float persistently in puddles. Ice melts into opaque white shadows. The city, gardens, lawns, trees look as if they've been steeped and soaked, stripped of life, chilled and pressed by the weight of snow. The surroundings, in their muddied bleakness, look as if nothing will ever grow again. Our eyes, accustomed to snow and dazzle, have to readjust to dun and grey and brown.

I look out of the window and the garden seems wrecked. Everything is puddled, laked, all liquid mud floating with a scum of the faintest ice remnants of white, lurgied and dead. Nothing will ever grow again. When it is warmer, brighter, more like spring, I'll collect up all the sharp needles of smashed slate – bits of last autumn's scaffolding, put up while the windows were being painted – from the stones. I'll clear the grass of branches broken and blown by snow and wind. The sturdy clematis by the study window is dead, and the roses are dead. Now I know the power of cold and the weight of snow. The *Cordyline australis* at the end of the garden which I didn't plant and never much liked but which thrived alarmingly behind the pond, sheds its long and spear-like leaves. Over the weeks, it begins to sag and pale and bend until its leaves hang and its thick, hemp-like stem topples, angled like an awkward elbow. (I should have wrapped it up and protected it from the cold.) Not just mine – all over the city, *Cordyline australis* has been afflicted by the prolonged cold. It is an unsuitable plant for the climate. I walk down the back lanes, and everywhere in gardens dead and dying plants are at one stage or another of their inevitable collapse. Over walls, central stalks point leafless like amputated stumps, neglected there in their death throes and after. They'll probably be there until time pulls them back into the fabric of the earth. As soon as I can get out to work in the garden, I'll saw mine down and by spring, the place where it grew will be covered by self-seeded bluebells.

February 9th

This morning I hear the voice of an oystercatcher, the first of the year. *Haematopus ostralegus* has returned from points south where it has wisely spent the winter. (Not that far south. Aberdeen's oystercatchers probably winter in the north of England and Wales.) Now, their inimitable high piping call will be everywhere, their busy presence, too, their air of ferocious concentration as they feed and fly and nest all over the city. *Sea-pie, sea-pyot, mussel picker* . . .

Oystercatchers have moved inland in numbers since the first record of their nesting here in 1966. Over the years since, they've established themselves until now they're almost the symbol of the city. There are more urban roof-nesting oystercatchers here than anywhere else in Europe. The building style of post-oil Aberdeen may have attracted them by providing suitable places to nest on the flat roofs of the new, modern buildings throughout the city – the roofs of oil-company headquarters, of the hospital at Foresterhill, of schools, flats, anywhere where there's an appropriate area and a small amount of covering gravel which can be scraped together to form a nest. (When some of the flat roofs of the university were renovated, special provision was made for the oystercatchers and suitably designed trays of pebbles were provided for nesting purposes.) On every patch of grassland, on every traffic island there's an oystercatcher, bold in its lovely outfit, this coastal wader, dipping and drumming its brilliant orange beak to pick at what's there to be eaten. They nest on the ground too, on the grass of the gardens at the castles of Drum and Crathes, in the grounds of Gray's School of Art.

The oystercatchers call a lovely, hasty, crazy, frantic call, trailing it wildly across the skies. They are commonly regarded as Gaelic speakers, with their cry of *'Bi Glic! Bi Glic!'*, 'Be wise, be wise.' (Prosaically enough, it sounds to me like 'eep eep eep'.) Oystercatchers have particular connections with the Western Isles, where they're known as *Gille-Brighde*, Bridget's guide, sent by St Bridget to guide sailors to safety. It was the bird who, during a legendary and otherwise unrecorded visit to Scotland, helped Christ hide under seaweed from enemies, for which service they were given a white cross to wear as a reward. The poet John Heath Stubbs wrote of it:

> But the sea-pies, flying,
> About the limpet-covered reef, with clear bright calls
> Took pity on him there, and in their scarlet beaks,
> Brought kelp and tangle to cover him . . .

Oystercatchers, *Servant of Bride, guide of Bride*. It's around St Bridget's Day that you begin to hear them call. It is 1st February, which is Imbolc too, the Celtic celebration of the lengthening of days, the midpoint between the Winter Solstice and the vernal equinox. Now they're back, I'll hear them every evening calling in the darkness as they fly past the house, as if they're singing their joyous, inebriated way home from the pub.

February 19th

A morning that's cold even for late February. I've put wood into the kitchen stove and broken the ice sheet on the outside birds' water bowl. I watch two wood pigeons chasing one another, lumberingly amorous, across the grass. I fill the metal baking tray from which the larger birds feed with bird mix, and top up the feeders for the smaller ones. I try to tidy up after them, knowing that if I don't the rats who, from time to time, move into the space under the house, will come again. I watch, but they haven't yet and soon, when the weather is warmer, when the prospect of snow is past for the year, I will rig up the devilishly clever rat-proof bird-feeding system upon whose design I have been working, so far unsuccessfully, for a long time. (Rats are far more ingenious than I am.) Collared doves perch on the branches outside the windows of the house, storey by storey. I hear the sound of their wingbeats through the glass then the shrill shriek of their flight call as they head towards the food tray. The witch hazel by the front door has bloomed in spiky yellow flower. In the cold sunlight, wood pigeons stop their wooing to tramp their heavy, stolid way across the stones to feed. They always know long before I do, as Chicken does, that, unlikely though it still seems, one day it will be spring.

Fellow Travellers

It's while I'm doing a bit of desultory garden-tidying on a fine cold day that I notice a small disturbance in the heap of last year's leaves still gathered in the corner under the kitchen window. It's as if they've been deliberately scuffed up. When I look closely, I see they have. There are signs of digging. My attempts at tidy bird-feeding have failed, and now there are rats.

There's a space under this house, much shallower than a basement, with just sufficient room to crawl should you be so inclined, if you're in training for potholing or intent on becoming a commando. The space is commodious, even luxurious if you're one of the smaller species of rodent. It's undisturbed and dark, and has an abundance of phone lines, electricity cables, beams, stones, rubble and the detritus left by builders many decades ago. It has provided, on a few occasions during the time we've lived here, shelter and sanctuary for wild rats. Their arrival appears to be closely related to the activities of those who, for purposes unknown, dig up roads. Removed from the security of their homes elsewhere, rats move in under my house. I've learned what to look out for: evidence of rat nibbling on the bird-feeders, or a little heap of soil or the removal of dry leaves leading to a run – a neat sloping tunnel that disappears steeply under the house wall. (Deep foundations and the formidable qualities of granite present no obstacles.) It's not just my lamentable failure as an inventor that is to blame – after birds have been about, there's always food scattered among the stones, unnoticed in the darkness. Besides, the homeless will, understandably, always seek a home. Having found evidence of residence, I'll watch and inevitably see a rat appearing hesitantly round the corner of the house. It will be a smallish rat, a brown rat, *Rattus norvegicus*. Equally inevitably, it will be an attractive, peaceable-looking rat, but a rat all the same.

Finding that *Rattus norvegicus* has established a headquarters under the house is a testing time for those of us who have a complicated relationship with rats, and while I don't share the commonly held view of them, I recognise my obligations towards my neighbours and society at large. I wish there was something I could do other than the drastic measures I'll have to take. I wish that aversion therapies worked, that peppermint oil or garlic or chilli sauce would drive them away, prevent them from breeding, stop them from establishing a colony. (Knowing that leaving rats to establish a stable level of population would be both the most far-sighted and the soundest course is not one I can easily argue with those who might be temporarily discommoded by my resident population. It's a remedy that would involve a more widespread plan of action than I can implement alone. Nonetheless, in relation to virtually every conflict with other species, it's probably the single most biologically sound answer.) The only thing to be done is to make the obligatory phone call to the council pest-control department. By now, after years of albeit infrequent rodent incursions, I am friends with Craig and Dod, the rat men. They are helpful, charming, appreciate the finer qualities of rats, do what they do in a spirit more of regret than conquest and bring presents of rubber mice for the birds to play with. I dislike the idea of using poison against any creature. I dislike the effects and the cruelty involved but there's no alternative I know of, no effective piece of equipment, no humane trap, no sonic device, nothing else that I can do to deter rats from living underneath my house.

As with so many other species, it's the language we use of rats that commonly limits our imaginative response to them, our own understanding of their circumstances and behaviour and the ways in which their world intersects with our own. For a long time, I've objected to the use of negative terms to describe crows and magpies such as evil or sinister, words containing an overwhelming weight of misplaced moral judgement. (They are birds, and therefore beyond our terms of morality. Besides, who are we to criticise anything on a species-wide scale on moral grounds?) I've tried to defend corvids against ignorance and unfounded prejudice. I've talked as widely as I can about their qualities, their brilliance, their place in schemes of ecology and evolution but attempting the same with rats is on a different scale of endeavour. The word 'rat' is in itself enough to bring about extraordinary reactions, shuddering revulsion, distaste, fear and only rarely appreciation of their adaptability, their intelligence, charm or beauty. Append the word 'rat' to anything or anyone, and the judgement is made, often with a vast imbalance of cultural, mythological and psychological history to prevent even the smallest degree of objectivity.

'Rat', in any case, is not a neat phylogenetic classification – it's a catch-all word used for many creatures, including ones who may or may not be of the genus *Rattus*. The order Rodentia, of which rats are members, forms the largest order of mammals – over 40 per cent, represented by 2,300 or so species. There are 51 species in the genus *Rattus* alone. Intelligent, successful and interesting as rats may be, they're creatures who, in moving from their origins in time and place, have proved unfortunate in their allegiances and associations, in their parasites and their commensal partners – those with whom they live in a degree of conjunction.

Like us and indeed with us, rats have travelled far in both space and time, in their case from their Palaeocene and Eocene origins 54 million years ago, from their anagalid ancestors, forerunners of both rodent and lagomorph – rabbit – families. Both *Rattus norvegicus*, the brown rat, and *Rattus rattus*, the black, have their origins in Asia; the former in Mongolia or China, the latter farther south in what is now Malaysia. Both began their worldwide spread as humans did, following trade routes. The black rat was first – evidence of *Rattus rattus* habitation in southern Europe dates back to between the fourth and second centuries BC, followed later by *Rattus norvegicus* (who, in spite of the name, don't have anything to do with Norway) although the exact date is still uncertain. From maps of the spread of rat populations, it's possible to trace the history of worldwide trade as rat travel mirrored the bold, innovative movement of humans across the globe, following river and sea routes, man and *Rattus*, pioneering travellers together. *Rattus rattus* arrived in the New World in the sixteenth century, *Rattus norvegicus* in the mid-eighteenth century, carried on ships from Europe.

Adaptable, omnivorous and hardy, rats thrived in most of the places they reached, establishing their reputations by their need and capacity for food, by the speed and success of their reproduction, their ability to gnaw, to eat and to destroy, although it wasn't until the end of the nineteenth century that the connection would be made between rats and plague.

Paradoxically, rats' biological fortune is also their misfortune, and lies in their ability to multiply. The sums are breathtaking – up to five litters a year, a gestation period of only 21 days, litters of up to 14. Rats' fecundity is their tragedy, the force and tragedy of life itself, the other side of the urge to reproduce. It damns them while giving them the power that lies in overwhelming, threatening numbers; numbers to induce our terror and our fear, to mirror in their ability to destroy, our own. (In spite of this, research evidence suggests that in Britain, estimates of rat numbers are greatly exaggerated and rat populations will probably decline as rainfall increases and causes the regular flooding of sewers.)

Anecdotally and in urban myth-making, there is a rat only feet away from any one of us at any given moment, the number of feet depending on how scared the source of this myth requires us to be. It's almost as if as humans, we have a right not to be as close to a rat but if we're near a rat, a rat is near us, its major source of danger.

Craig and Dod arrive. We have our usual conversation about the finer qualities of *Rattus norvegicus*, full of wistful regret (I ask them again what known alternatives there are to poison although we all know that there are none), and then the poison is placed. I try not to think about the consequences; about the pain I have caused to be inflicted. I wonder too, what is it that frightens us about creatures which seem to have the power to threaten by their forceful abundance? Perhaps it's our correct and increasingly cogent fear of the power of nature, the thought of forces beyond our control, unknowable and, worse, unstoppable. Rats seem to represent our worst fears as we concentrate the darkness of the world in the persona of one small creature. Rats have, throughout history, been seen as representing a level of depravity which seems to give us licence to use them as symbols of behaviour better reflecting ourselves than them. Just as crows were regarded with horror after the Great Fire of London when, entirely appropriately and practically for both crow and human, they fed on corpses, during the First World War rats were particularly loathed because of their presence in the trenches, as they attempted to feed on the unfortunates we ourselves had slaughtered.

When it comes to the natural world, it's difficult to escape the long shadow of ideas. Many of our attitudes are still profoundly influenced by Plato's and Aristotle's concept of a 'Scala Naturae' or 'Great Chain of Being', an imagined hierarchy of creation which for centuries determined all consideration of and behaviour towards other species. Added to and reinforced by later philosophers and theologians – St Augustine, Thomas Aquinas, Descartes and the thirteenth-century Spanish Franciscan Ramon Llull among them, creation was systematised into a form often portrayed as a tree with God at the top, descending by degrees from the realm of the spiritual as represented by angels to that of the inanimate, with all life forms arranged in between. Implicit in the system was a scale of diminishing value and although below angels, humans were still in sufficiently lofty a position to feel that they could behave as they chose towards all other species, reassured in this belief by the accompanying set of self-serving assumptions made about the mechanistic, feeling-less lives of animals. When the seventeenth-century Catholic theologian Nicolas Malebranche wrote: 'Animals eat without pleasure, cry without pain, grow without knowing it. They desire nothing, fear nothing, know nothing', it is not difficult to appreciate the degree of licence his assertion gave to humanity.

We perceive rats as being dirty but, like birds, they are clean to the point of obsession. Condemned as disease carriers, they are only as we all are or might be, as many other animals and birds are, ones who may have less of a role in the grimoires of our horrors. The accusation often seems to provide justification for removing creatures, not only rats, from places where they may be doing no harm, and for acting towards them with unwarranted cruelty. Everything spreads disease, given the correct circumstances, particularly now, in a world of wind and air and water and mass travel, of import and export, of grotesque and unfortunate farming practices. Unknown numbers of pathogens seep and infiltrate and blow through the ventilation systems of aircraft, through ballast water, on mud carried on boots and by every other strange and unlikely method: they have the means to threaten and infest the world. Many diseases are carried by other species just as they are by the human carriers of viruses and bacteria (although appearing to be involved – even in some generally passive way, as vector or carrier – in spreading plague does no good at all for one's reputation). Even that connection has recently been challenged by the archaeologist Barney Sloane in his book *The Black Death in London*, in which he suggests that the absence of rat remains from the fourteenth-century epidemic, as well as evidence of the speed of the spread of disease, indicate that transmission was person to person. (Besides, humans have little to be proud of – the bacterium that causes plague, *Yersinia pestis*, has been both widely investigated and even used as an agent in biological warfare.)

The perception of the rat in myth and culture is far from straightforward. Admired in some cultures for their qualities, their resourcefulness and success as a species, they are equally reviled for their destructiveness. Like many creatures, they fare rather better in Eastern cultures, where, sometimes being regarded as sacred or as bringers of fortune, they are sympathetically portrayed in painting and sculpture. In Chinese horoscopes, those born in the Year of the Rat are considered charming, imaginative, sociable and a host of qualities that may or may not have any relationship to the perceived behaviour and habits of rats. The Temple of Karni Mata at Deshnok in Rajasthan is devoted to the rat, which is deemed sacred within its precincts. Rats in many cultures are thought to be agents of revenge or potential rivals, in stories such as that of Bishop Hatto being eaten by rats as punishment for food-hoarding, or in the many fantasies in which rats replace humans as the most powerful species on earth.

In the early 1980s, the French graffiti artist Blek Le Rat drew rats all over Paris, to honour the rat as 'the only wild animal living in a city'. In London, Banksy portrays the rat as an urban character and hero, or anti-hero, as anarchist or gangster; rats easily scaling walls daubed with warnings of anti-climb paint.

As is the fate of many creatures supposed by humanity to have no place in society, the name and image of rats has been co-opted into the language of hatred and prejudice, as descriptive of what is, in fact, the worst of human behaviour. Boria Sax's masterly book *Animals in the Third Reich* examines the often convoluted and contradictory attitudes of the Nazis towards the natural world, their selective extension of protections, their notions of purity and favoured species, their equation of some, including rats, to those held to be racially inferior.

In his poem 'Burbank with a Baedeker: Bleistein with a Cigar', T. S. Eliot writes of the rats underneath the piles of the Rialto, and of 'the jew' being underneath the rats and while much discussion has ensued over the years about Eliot's intentions and motives in writing this, my only response, having read an amount suitable to inform myself as fully as I wish to be informed on the subject is – *whatever*.

Craig and Dod leave, promising to come back again in a few days to check. As the poison does its work, I think of my own long relationship with rats, of the years since we bought our first one in a London pet shop, not only the first rat to share our house but the first creature of the many who would. A splendid specimen of the species, a 'fancy' rat, Rupert was everything one might require in a small, easily portable, intelligent, interesting, affectionate pet. He was beautiful too; soft and grey-brown of fur, delicate and graceful as rats are in form and movement, in the gloss of their coat, in their neat and dexterous hands, pink ears and fine, sensitive whiskers, the perfect netsuke creature. His tail, as are the tails of all rats – the cause of much human revulsion – was a wonder of elegance and efficiency, an instrument of perfect balance, a sensor and measuring device. He spent his given thousand days happily, I hope, permanently in our company. When he died, we bought two small females without knowing that one of them was already pregnant. By the time her litter was born, we had learned sufficient biology to realise that we had to separate even very young males and females quickly. We became adept at rat-sexing – a slightly less simple procedure than it sounds, the genitals of male and female infant rats looking remarkably similar, and one which requires at least two rats for comparison, ideally one of each sex. (A book on rats helpfully provided photographs.)

An almost silent presence in our noisy household of bird calls, rats were with us for many years, contributing as they could to the general melee with their repertoire of squeaks: their play squeak or their mutual-grooming squeak, their combative squeak, their squeak of pain or the squeak that's an extended yittering made in response to fear. There was the occasional hiss in circumstances of danger, and the many moments of satisfaction when they'd crunch their teeth softly in what we hoped was rat bliss. (Some report having heard rats laugh in response to tickling, but they're all sounds you're unlikely to hear from a rat unless you're up pretty close and personal.)

The 'fancy', the process that allowed us to have such relationships with rats, is selective breeding whose origins lie in the rat-catching and rat-baiting of eighteenth- and nineteenth-century Britain when breeding to encourage qualities of strength or ferocity was undertaken. The practice expanded to encompass breeding for colour and other features, initiating the separation of rats into the wild and the domesticated. At the beginning of the twentieth century, a lady called Mary Douglas (described as 'the mother of the rat fancy') sought permission to show rats at a local mouse show, after which rats became increasingly popular pets. Being bred for qualities other than good health, they are subject to diseases, tumours, breathing difficulties, heart disease and even epilepsy, afflictions that bedevil and shorten their already short lives, but it is 'the fancy' that has made it possible for humans to treat rats as any other small and delightful pets while all the time we know that, whatever their colour or breed, whether they're blue agoutis or lilac agoutis, cinnamon pearls or silver roans, black-eyed Himalayans or Russian blue point Siamese, they're all *Rattus norvegicus* in posh fur coats.

Not long ago, a relative who is a politician sent me a letter he had received from someone who runs a pet sanctuary urging him to find a way to limit the sale of small animals, as the sanctuary can no longer deal with the numbers being brought to them. Reading it reminded me not only of my years with rats, but of the years when I used to look after the pets belonging to the nurseries the girls attended, during the summer; of the responsibility of taking care of other people's pets, of hamster- , gerbil- or guinea-pig sitting and the involvements we have all had with sundry creatures at particular times in our lives. Often, we buy them for our children as lessons in mortality but who wants to be a lesson in mortality? (In the course of time, the lesson seems all too often to be the lesson of relief when the obligatory cleaning and feeding is at an end, the only shock, or guilt, being the brevity or absence of grief.) We want to respond to our children's interest, to encourage responsibility, nurturing, a love of the natural but is the natural world best represented by the creatures we buy? We keep reclusive desert animals, or naturally secretive ones and expect them to be sociable. Diurnal as we are, we keep nocturnal or crepuscular pets and are frustrated when they respond, as they have to do, to the rhythms of their own lives, with their night-time rustlings and daytime somnolence. Hamsters seem to fare badly – mention the very word 'hamster' in virtually any company and there'll follow a confessional roll call of brief lives and often horrible deaths, of disappearances into gas fires, through holes in floors, of injuries worthy of a journal of trauma surgery: lost limbs, broken spines, missing ears. I think of an account of a pet rabbit forgotten, found starved to death alone in the garden shed. There are the escapes, too; flat-dwelling gerbils, hamsters, mice who disappear, never to be seen again. Thinking of these small losses, or valiant bids for independence, I imagine a place like Mary King's Close in Edinburgh, a subterranean maze of ghostly seventeenth-century streets, preserved after the High Street was built over their foundations, and think of a world of nocturnal sounds, matings, unseen multiplyings, where generations sired by escaped and neglected pets slowly, inexorably bring about an unassailable, numerical vengeance on a world that too easily acquires delicate living things.

During the long-ago youth of our rats, a colleague of David's who was visiting us, on catching sight of them in their houses in what we have always called 'the rat room', wandered off to admire them. He asked if he could hold one and then spent the evening chatting to them, allowing them to hide in his pockets or sit whiffling eagerly on his shoulder. Not known as a sentimentalist, he was a scientist, involved in work with the most pressing and urgent neurological problems of the time. At the end of the evening, enchanted I think by the earnest, concentrating stare with which rats will engage you (a stare dictated most probably by deficiencies of eyesight), he said, 'I wonder if it's really necessary to experiment on rats?'

For 200 years and probably more, rats have been the favoured animals for laboratory experimentation. From the first to be bred specially, the 'Wistar' rat, named after the institute where work was begun in the early years of the twentieth century, multiple strains have been bred. There are Sprague-Dawley rats and Zucker rats, Long-Evans rats, cloned rats and transgenic rats, and sometimes it seems as if we demonstrate our full power of the subjugation of another species by the range of purposes to which we subject it, from the scientifically ambitious to the abjectly trivial. Rats are used because they're plentiful, amenable, suitable for a wide range of experiments and cheap to maintain. Over recent years, they have been used increasingly in research in genomics – the study of the genetic heredity of an organism. When the genome sequence of *Rattus norvegicus* was announced in April 2004, it was one of only three mammals – rats, mice and humans – to have been sequenced, and while in this and in other ways the contribution of rats (and many other creatures) to medical science has been vast, my friend's words have stayed with me, resonating over the decades, although it is the reasons for and methods of experimentation that leave most questions in my mind.

Observing rats over many years impressed upon me the complexities of their social behaviour, their capacity for empathy, reciprocity and cooperation, all behaviours that have been subjected to extensive research.

We watched our two young female rats getting ready for one of them to give birth. The pregnant rat chose to give birth inside the large glass sweet jar we had put in their house, and both prepared by tearing newspaper and lining the jar to make a nest. After the birth, while the mother looked after the young, the other rat cleaned, removing the dirty newspaper by pushing it to the opening of the jar, licking, cleaning the jar and replacing the paper, something she did regularly. Both rats appeared to raise the young, exercising discipline and carrying them from place to place. We watched sick rats being tended, stroked, groomed by their siblings, and their evident distress at the death of one of their number.

In *Wild Justice: The Moral Lives of Animals*, Marc Bekoff and Jessica Pierce quote from Russell Church's 1959 paper published in the *Journal of Comparative and Physiological Psychology*, 'Emotional Reactions of Rats to the Pain of Others', in which it was demonstrated that rats will not take food if their doing so causes pain to other rats, and that they will help other rats in distress – the motivation for doing so probably being empathy. Pointing out the irony of the fact that studies of empathy are often ones of the most exemplary cruelty involving inflicting pain and death within sight of other animals, Bekoff and Pierce suggest that good evolutionary biology of itself should indicate that empathy exists in other species. (To whom is it a surprise to learn that rats display signs of distress when obliged to witness the decapitation of other rats beside them?)

In a 2006 paper called 'Generalised Reciprocity', Claudia Rutte and Michael Taborsky of the University of Berne demonstrate that rats who have been previously helped by other rats are more likely to offer help themselves, while the results of a study undertaken at the University of Chicago published in 2011 show that rats, on being faced with two cages, one containing chocolate and one another rat, will quickly learn how to liberate their fellow before opening the second cage to share out the chocolate. The benefits bestowed upon the group by the cooperation of individuals is clear. In *Wild Justice*, Marc Bekoff quotes Martin Nowak, director of the Program for Evolutionary Dynamics at Harvard University: 'Cooperation is the secret behind the open-endedness of the evolutionary process. Perhaps the most remarkable aspect of evolution is its ability to generate cooperation in a competitive world.'

For many years now, long-held beliefs about the limited nature of bird and animal cognition have been undergoing a process of revision as a result of neuroanatomical and behavioural research. At Cold Spring Harbor Laboratory in New York State, where researchers have established a 'rat cognition movement' which uses rats instead of primates to study cognitive processes, recent studies have shown that rats presented with different types of visual and auditory stimuli are able to process and react to the information as efficiently as humans, results that might call into question aspects of work into the cellular basis of depression being done at the same institution. This work involves the use of the notorious 'force swim test' (also called the 'behavioural despair test'), in which rats are forced to swim in containers from which they cannot escape, and the euphemistically named 'learned helplessness' tests in which animals are subjected to various sorts of suffering in order to test their level of determination. The questions remain, and I cannot answer them. If rats are cognitively able, how far is it allowable to subject them to the kind of research methods prohibited for use with primates? Where are the limits and boundaries of what, as one species, we may do to another? Should the potential intelligence of our subjects limit or affect what we do? What do we mean by intelligence? How do we measure it? Since all our tests are human constructs, do they impose constraints on our understanding, or on the behaviour of our subjects? How far might our pre-existing assumptions affect the results of our research? What do we know of sentience and consciousness? How can we balance the needs of humankind against the moral questions of how we act towards other species? What is changed by our appreciation of their suffering?

The last of the rats we kept we had while Bec and Han were still at home. He was a retired lab rat given to Han by someone who had worked with him and felt sorry for him after his lab role was over. He was a young albino *norvegicus* with the blurred, dichromatic vision of most rats, slightly matted fur and a torn ear where his lab tag had been removed. We didn't know what he had experienced or what his contribution to the world of science had been, but he was the only rat we ever encountered who was unable to clean himself, as if his rat instincts had been oblated, whether by experience or breeding we couldn't know. He lacked the personal fastidiousness of rats and had to be helped to groom and to wash. (Lone rats may suffer in this way – it may not have been his previous life but his being alone and lack of a washing companion that prevented him from learning or maintaining the usual rat standards.) He seemed, for all his gentle and willing charm, a diminished creature, lacking the essential rat-ness of the others, the spark of vitality and purpose. It wasn't a surprise when he died quite suddenly in what was still his youth. Han had named him Elmo.

With Elmo, the moment and rats moved on. The girls would be leaving home soon and besides, there were enough birds to occupy space and time.

In adulthood, Bec continues the tradition and keeps rats. First, there are Alice, Cora and Celeste, and after the requisite time elapses, Greta and Selina. Usually they're kept in twos or threes, but Greta dies young and Selina is alone. She is a pretty white rescue rat. Being alone gives her unusual freedom as well as placing an obligation upon her human companions to make up for the lack of rat company. As a result, Selina has been known to make her way around Edinburgh in the privacy of a handbag. It is possible to buy rat harnesses, which, in theory, allow you to take your rat for a walk. Encouraging the rat to wear the harness is only the first of the potential difficulties.

Selina is allowed to roam freely round Bec's flat, which has been accomplished (up till now) without mishap to anyone and particularly since the advent of Wi-Fi, without significant and expensive damage to Ethernet cables, which, with the unerring instinct of the rat, used to be the invariable object of desire. Who would have thought that, in terms of avoiding incarceration as punishment, advanced wireless technology would be such a boon to the rodent freethinker?

Selina arrives with the family for Christmas. The time becomes a contest of wills between us although she is a rat of whom I had previously suspected no particular strength of character. On one afternoon, after we have been playing with her, I shut her into her house in a room on the first floor. When I see her next, she is walking calmly upstairs from the ground floor. The door of her commodious holiday accommodation is open, and I worry that, inadvertently, I left it that way. I shut her in and close the door firmly, remaining confidently certain of her incarceration until the next time I see her when she's strolling across the landing. This time when I replace her, she's angry. She grasps the bars with her hands and shakes them with frightening strength and vigour. I twist wire around the door to keep it shut. I know that she's used to freedom and to wider, more exciting possibilities than she will encounter here – Edinburgh, the capital city, I seem to recall, is such a place. I reassure Selina as I place her back again, that her visit to the provinces will soon be over.

This goes on for days. She escapes. I find her, eventually, and put her back in her house. There is no – or little – hardship involved for her. Her house, were it mine, would be, relative to our sizes, Holyrood Palace. On every occasion I tell her not to worry, that soon she'll be home. Again and again I try to find ways to prevent her from escaping. I tie her door. I wire her door. I move her house against a wall. I pile books against her rewired door. I go away then return some time later to check on her, but she has escaped again. The wire lies chewed on the ground. As for the books, I can only assume that she bypassed them in some way unimaginable to bony, large, unwieldy humans who can't mould themselves flat, as rats can. On the final day of the Christmas holiday, she escapes again. I find her like the princess from 'The Princess and the Pea', lying between the top layers of a neat pile of duvets on the bed in the spare room. She is asleep, pink hands folded. I leave her there and close the door. Later in the day, she's woken, picked up and placed in her travel house to be borne off back to her even more majestic home in the capital, the one in which she spends remarkably little time; three storeys of rat bling, rat hammock, rat tunnels, lavatory area, mezzanine floor and all. I do not lie and say that I'm sorry to see her go.

Naturally, during her visits, I pay close attention to Selina's gustatory preferences. She's less fussy than many rats I've known but nonetheless I think of her as I listen to a question on a gardening programme about whether or not planting mint deters rats. The answer involves wide generalisations about the tastes of rats. In my experience of rats, or fancy ones anyway, they do not share tastes but then perhaps 'the fancy' has removed rats' universal tastes, given them specialised ones for ciabatta and olives, for strawberries, avocado, basil, bananas and particular brands of dark chocolate, while inuring them to the charms of Twiglets or Marmite. Perhaps it has corrupted them so far that now, they are as subject as we are to questionable, spurious, mercantile notions of 'choice'.

The poison must have worked. Craig and Dod visit a couple of times to check. There are no more rats in the garden, or under the house. I hate both the knowledge and the suffering, and am aware that at least part of my feeling is that the view I have of myself as benign, a lover of all creatures, is patently undermined by this act of proxy killing. I think of the very fine bridge between one state of being and another, from the state of the wild to the state of chosen pet or the favoured subject of experiments. I accept that we live in a world of contradictions, although it makes me feel no better.

WINTER INTO SPRING

March 1st

I go into the garden to assess the winter's damage. I'll have to cut down and stack everything that's dead or broken. I was wrong about the effects of cold and snow. A shoot of green is growing from the apparently dead tangle of stems – the clematis lives! The pond is still covered with that opaque film that's always left by melted ice.

I'm looking through some gardening books for advice on caring for clematis affected by cold, when I come across the gardening notebook I kept when we first came here. It's very neat, with planting plans, plant names and seed merchants' details (when was I ever so organised?). In it, I find a photo. A little girl, smiling, is running through a garden, burdened with drawing books and pencils and by the trailing tartan picnic blanket clutched under her arm. The garden behind her is walled, not large, though large enough for the centre of a city. It has a bleak look, almost institutional, raw and open to the sky. There are only a few low shrubs, small and clearly newly planted, isolated in the expanses of flowerbeds. It needs plants, flowers, trees. A few beech branches from the tree in the next-door garden reach across the wall but the rest of the tree can't be seen. Nor can the worst of the garden's neglect, much of which has already been dealt with, broken stones and browning, dying, crowded trees cut down and carried away, their roots dug up and removed as thoroughly as roots can be removed, old rubbish and dumped things dug out of every border, paths relaid and new fresh soil put down. A doo'cot, a converted coal shed, waits for inhabitants. The garden faces east, towards the sea, a line of white in the distance beyond the walls and city. It's a garden in waiting. Things will be planted. All it needs is time.

This garden was a challenge. Our first garden was in perpetually raining Lochaber and our second in London. The former was over-grown and neglected, the latter just neglected. I'd never faced a blank and vacant space.

The notebook has lists of the plants I chose. The first were old roses with French names, the name of statesmen and queens, wives and cour-tesans. I read the seductive descriptions of each, of centifolias and moss roses (the moss roses, Alfred de Dalmas and General Kleber, with their scent of balsam and sweetness), damasks and bourbons, the climbers and the ramblers. There were too, the Königin von Dänemark ('Raised in 1816 by Booth and introduced in 1826, the finest of the Albas'), Alba Maxima, the Jacobite rose ('They open with a suspicion of pink'), Mme Alfred Carrière ('often to be seen growing on the walls of old houses'), Adelaide d'Orléans, Félicité et Perpétue, Paul's Himalayan Musk. They arrived in damp packages, carrying their history and habit, *remontant, climbing, summer flowering, perfumed*, everything they would become, there in a pile of bare brown twigs.

I read books and prepared as I should, dug in everything particular plants were said to require, and hoped. The plants arrived or were bought as small stems in plastic pots that seemed to overwhelm them. Even more when I planted them out – they looked tiny, swamped by the earth in which they were to grow, little islands in the emptiness of the flowerbeds, jasmines and clematis, buddleia and lavender, honeysuckles, viburnums, lilac and philadelphus. I planted *Hydrangea petiolaris* and ivy, an espaliered apple and pear along the back wall, a forsythia and small trees, a plum, a crab apple, a little rowan for luck. A friend from Lochaber and I sang a chorus of 'Rowan Tree' as I planted:

> Thy leaves were aye the first of spring
> Thy flowers summer's pride,
> There was nae sic a bonnie tree,
> In a' the countryside.
> Oh rowan tree . . .

The more hideous bits of statuary that were in the garden when we took it over, we gave away to someone who takes away hideous garden statuary. The rest, the almost acceptable, we kept. Time has transmogrified them, mellowed them to the classic, the moss-grown and gentle, the square bird bath on its plinth, the coy lady by the back gate, the child and the dolphin, weathered and beautiful now, as if they're part of time. The odd little squirrel and owl, spared because the children liked them, stand, small sentinels on the back step, no longer odd, but part of the years. Among the ferns, there is the sculpture of a butterfly resting on an open hand which David carved from sandstone, and under the maple his lovely carving of a lady with a rook on her shoulder.

Not long after the garden had been first planted, the unseen mouthparts of one creature or another applied themselves to the leaves of a newly growing plant although I don't remember which, or what. Dutifully, in the spirit of defence and outrage at this depredation, I went to seek the advice of a man at the garden centre who looked as if he might well have some old-time, old-world advice. Instead, he led me to a shelf and offered me a spray. 'This', he said, 'will kill everything that moves.' I didn't want to kill everything that moved. I didn't, when I considered it, want to kill anything, ambulant or otherwise. Did holes in leaves matter? I put the spray back on the shelf among the other sprays and cans and bottles (many of which will, by now, have been banned) and went home and for the next two decades and more, have used nothing to remove anything from any plant or leaf, nothing to deter or to persuade. If I find coral spot, which occasionally I do, I cut the few small speckled branches off and burn them in the stove and hope it'll go away and it does, who knows how. I put down compost and seaweed mixes and after their deaths, suitably marked and mourned, dig in the bodies of the birds and creatures who have enlivened the life of the house and will now enliven the life of the garden. Most of my gardening policy adheres to the principle of *wu wei*, masterly inactivity, or more honestly, just inactivity.

We installed the doves we had been given in the doo'cot, and after their initial period of acclimatisation during which they had to be shut in, we let them out to fly and soar over the roofs, the garden, the neighbourhood. I'd pass them as I walked home up the lane, a wide turning circle above me, a swirl of vibrant white. It seemed strange but wonderful, a privilege to claim acquaintance.

The scruffy grass was removed and new turf laid. It grew well enough although none of the potions usually applied to keep lawns green and weed-free and perfect ever were because the doves pecked on the grass, and other birds too. I cut it, raked it occasionally and did little else, but still the grass grew and grows, or rather, the moss does. In spring birds, particularly blackbirds, scratch it up and carry it away for nesting. The rain has increased over the years and with it, moss has invaded the north side of the already far from perfect lawn.

How little faith I must have had in growth, or in my own involvement in the processes of growth and growing. I didn't imagine on the day I took the photo, what would be, what would grow and what wouldn't, how over time and years the garden would be what it is now, dense and high and green, a place where the *Hydrangea petiolaris* coats the walls so heavily that every summer, windows and doors have to be carved out of its growing leaves. I couldn't have seen that the small ivy on the east-facing wall would reach so ambitiously high, steadily climbing towards the top of the roof, regularly invading the gutters from where it has to be cut back, that the *Viburnum carlesii* and *Philadelphus coronarius* would form a wall of green in summer, a refuge for a population of diverse birds or that the beech tree would have reached the top windows of the house, spreading a light-defying canopy of leaves over the garden below. Everything grew. Now, the forsythia on the north wall tangles into quince and plum, yellow sprays of flowers, the first of late, dark winter falling over the snowdrops below. Ivy-leaved toadflax seeded itself firmly into the stones and trails its tiny purple orchid-like flowers delicately down the garden walls. It dances in summer with tiny bees. The little rowan, planted too close to the wall, has grown into a large rowan. Ivy spreads across the stones in the wall like a network of fine veins; stems of quince, *Chaenomeles japonica alba*, extend thickly over the top of the wall, intertwined like interlocking arms. The Paul's Himalayan Musk, whose delicate pink blossom belies its ferocious habit, has grown to a fairy-tale rose, the kind that provides passing princes with a challenge. (I take the saw to it from time to time, an activity more akin to wrestling than gardening.) The Adelaide d'Orléans at the front door swiftly grew dense, weaving inextricably through the *Wisteria sinensis*. David named it 'the swain trap', hoping it might deter the girls' eventual suitors. It didn't.

As the ivy grew on the back wall, a dove would occasionally nest there in preference to its house. I began to hang bird-feeders from branches, and squirrels came and a few birds, mainly blackbirds and robins and over the years, as the plants have grown and filled the empty spaces, the bird population has changed in species and in number and the squirrel numbers have diminished.

What I planted has flourished far beyond what I could have imagined on that day when I framed a broad sky and bare walls in the viewfinder. Everything has changed and grown: the tree, the empty flowerbeds, the small girl on the grass.

March 3rd

The hellebores are beginning to flower in their corner under the branches of the beech. Their names, Christmas Rose (*Helleborus niger*) and Lenten Rose (*Helleborus orientalis*), seem to bear no relation to the times of their flowering because here they are, deep pink and pure white, flowering together. I can't remember if they always flowered at the same time but would know if I had noted dates and flowering times, and all the other significant moments in the natural life of the city and the garden, as people have done for the entire span of Man's life on earth. I wish I'd inscribed everything as people once did, on oracle bones in China in the Shang Dynasty 4,000 years ago, scratching out details of rainfall and millet harvests, or as in the Xia Xiao Zheng, 'the Small Calendar of Xia', a month-by-month commentary on the weather, plants, animals and stars of the Western Han Dynasty, 200 years BC, or as the ancient Egyptians did, and the Mesopotamians, as agricultural communities everywhere. It's what is now called phenology, the study of the timing of recurring natural events.

Observation of the natural world has always been a subject of study for philosophers and scholars – for Aristotle and Pliny the Elder and later for the tireless polymathic giants of the Renaissance, Konrad Gessner, Ulisse Aldrovandi and in the eighteenth century, Carl Linnaeus, all of whom laid the foundations for the later scientific study of phenology. (The term 'phenology' itself was first used in 1849 by the Belgian botanist Charles Morren in a lecture at the Academy of Brussels.)

Konrad Gessner, who was born in Zurich in 1516 and died of plague at 49, was a botanist, zoologist and mountaineer. In addition to compiling a compendium of all the authors who had ever written in Latin, Greek and Hebrew and an extensive work on languages, he wrote the *Historiae Animalium*, a 4,500-page work on animals and fossils, and *De Stirpium Collectione*, in which he gave one of the first accounts of the dates of the flowering and leafing of over 1,000 plants. (He was also the first to describe *Rattus norvegicus*.)

A botanist, Ulisse Aldrovandi was creator of one of the first botanical gardens in Europe and an avid collector of both plants and natural curiosities. Born in 1522 in Bologna, he wrote fourteen volumes of natural history, a compendium of monsters, the *Monstrorum Historia*, and the *Historia Serpentum et Draconum*, a history of serpents and dragons. It was Aldrovandi's work that led Carl Linnaeus to regard him as the founder of the subject of natural history. (His collection of curiosities, his *Wunderkammern*, was broken up after his death although the surviving portion is still kept in the Museo di Palazzo Poggi in Bologna.)

Linnaeus, whose life spanned much of the eighteenth century, not content with undertaking the massive work of classifying everything that lived and breathed (and some things that did neither, since he included stones), first published as the *Systema Naturae* in 1735, was fascinated by the timing of natural events. In *Philosophia Botanica*, he established the criteria for recording annual data and in the *Calendarium Florae*, published detailed observations of plant and climate phenology. At the same time, Robert Marsham, a Norfolk landowner born in 1708, the year after Linnaeus, was keeping records of annual natural events as he did throughout his long life, noting dates of first flowerings, leafings, the first time cuckoos and other birds were heard. Marsham began his recordings in 1736, refining his observations to his '27 Indications of Spring'. After his death in 1796, his family continued the work of observation until the middle of the twentieth century.

Gilbert White, another careful recorder of his surroundings and garden, corresponded with Marsham after the latter wrote to White on the publication of his best known work, *The Natural History and Antiquities of Selborne*, in 1788 and 1789. White also wrote *The Garden Kalendar*, *Flora Selborniensis* and *A Naturalist's Journal*, in which he recorded his daily observations from 1768 until 1793. The journal is minutely observed and charming:

Jan. 22 1784   Hard frost. Snipes come up the stream.

Sept. 5 1784   Vast dew. Sun, sultry. No wasps yet: no mushrooms appear. One *fly-catcher* at Faringdon. Annuals make a
great show. Heavy clouds about.

July 15 1787   Mr White of Newton finds mushrooms in his fir-
avenue. Tremella abound in my grass-walks. The wet season has
continued just a month this day. Dismal weather.

The twentieth and twenty-first centuries have provided no fewer out-standing phenologists than previous centuries although the subject itself has altered, gaining in urgency as changes in the climate affect the natural world. Richard Fitter, described as one of the most influential of urban naturalists, author of an astonishing number of authoritative books on birds, wild flowers, butterflies and all aspects of the natural world, kept important phenological records for fifty years of the twentieth century and, with his son Alastair, in a 2002 paper published in *Science* magazine, presented data based on their observations. The average first flowering of 385 Oxfordshire plants, they wrote, was happening 4.5 days earlier during the last decade of the twentieth century than in the years 1954 to 1989.

T. H. Sparks, Britain's foremost contemporary phenologist, initiated the UK Phenology Network – Nature's Calendar – in 1998, encouraging online recordings from members of the public, using the resources provided by earlier recorders such as White and Marsham to compare dates and timings. Writing with others in a paper published in the *Proceedings of the Royal Society, Biological Sciences* in 2010, 'A 250-year Index of First Flowering Dates', Sparks shows that flowering is occurring between 2 and 12 days earlier than it did 25 years ago. The figures correlate closely with temperature recordings, flowering taking place 5 days earlier with every one-degree rise in temperature. In a recent book edited by Norman Maclean, *Silent Summer: The State of Wildlife in Britain and Ireland* – a title which invokes that of Rachel Carson's percipient book *Silent Spring*, published in 1962 and regarded as one of the foundations of a new ecological awareness – Sparks and colleagues in their chapter 'Climate Change' discuss the effects of anthropogenic changes on Britain's climate and assess their possible influences on wildlife, such as alterations in the synchrony of food chains and relationships between species.

Russell Foster and Leon Kreitzman write in their book *Seasons of Life* about lost connections with nature, the profound awareness which made people appreciate every fine-tuned aspect of the world around them and give an account of the French explorer Samuel de Champlain who, on his arrival in Cape Cod in 1605, was informed by the Wampanoag people that the best time to plant corn was when the leaf of the white oak was the same size as the footprint of a red squirrel.

March 4th

I'm reading Sybille Bedford's *The Sudden View*, an account of a journey to Mexico in the 1950s. As long ago as that, she mentions 'heat islands', something I had thought a relatively new idea.

It's not difficult in the unrelenting grey of late winter to find the idea of this city being a 'heat island' improbable, and yet while it may not always feel very much like it, we are. We may be northern and cold but if we're in a city, we're in a heat island, a place where temperatures are higher than the surrounding countryside by as much as eleven degrees. We're heat islands because of the way we build and the materials we use to build – our concrete, glass and stone. It's our roads, our roofs, our pavements, our dark, absorbent surfaces, our paving of gardens, our cutting down of trees. It's the way we live too, our heating or our cooling, our determined use of cars. We're bounded by the heat we generate, our energy rising to the skies. The heat raises the levels of particulates in our air, makes clouds and smog to obliterate the sun. The clouds bring rain, which evaporates from our impermeable surfaces to create more heat. Stone cities like this one absorb the heat of day and release it during the night and while we rarely, if ever, experience the still heaviness of more southern cities, our surfaces are as bleak and heat-absorbing, the pace of our destruction of gardens and trees as relentless, our implacable determination to expand our road networks and render obsolete the ancient art of walking. The only difference between us and the cities where heat is felt and suffered is that we hardly notice it at all. It often feels as if – all Enlightenment prescience – the builders of Scottish cities anticipated future developments in CO_2 emissions, and with care and purpose built cold islands, designing our cities to feature ice pockets and urban tunnels where bitter winds are channelled into so concentrated a space that you're uncertain if you're being blown by the wind or selected for special persecution by it. The writer Lewis Grassic Gibbon described living in Aberdeen 'like living in a refrigerator'. Of Edinburgh, Robert Louis Stevenson wrote:

And meantime the wind whistles through the town as if it were an open meadow; and if you lie awake all night, you hear it shrieking and raving overhead with the noise of shipwrecks and falling houses . . . the look of a tavern or the thought of the warm, fire-lit study, is like the touch of land to one who has been long struggling with the seas.

94

Quite.

When I came to look for a house here first, an estate agent warned me against one particular street. 'You don't want to live there,' he said sepulchrally, 'it's a cold street.'

March 8th

Now that the snow has gone, it's like walking a different city. I hear the Denburn again, unfrozen, making its way through its subterranean channels underneath my feet. Even when there isn't snow, I drive only when I have to. I'd cycle but I'm a coward. Here, only the truly intrepid cycle, people who aren't scared of vehicles higher than their heads, cars like urban tanks or vehicles of war.

This is a city where you can walk easily from end to end. The edges of it spread but not too much. They meander out in a modest way in several directions into industrial parks that, if you stray into them, seem more like intricate mazes designed to keep you circling, lost for ever among their chain-link fences and variation-on-a-single-theme names but even from places like that, you're not shut in. You can see the sea from everywhere, from all the hills and granite coastal platforms the city is built on which made the site so difficult to settle. Once, it was bisected by streams, all bogland, all dips and sloughs and hollows, all hills, declivities and valleys, but they've been built on, turned to a topography of stone. When you're travelling from the south, approaching the city is like entering a 3-D map; from the brow of the hill, the entire city is spread below you; its life and history rises between rivers and the sea, grey stone and spires and water. The rock cliffs are behind you, south; to your right, unseen, beyond the spill of houses, the wide arc of sea, the broad flash of beach and dunes.

There aren't any vast motorways, only a few dual carriageways, busy but easily negotiable. Once you're out of the city, you drive on almost empty roads, specially the coast roads south that seem suspended in the light above the sea.

'Cities approach immortality while everything within them rises, falls, is erased and transformed and replaced,' Rebecca Solnit writes in her book, *Infinite City: A San Francisco Atlas*. In a city, it's impossible to forget that we live in places raised and built over time itself. The past is underneath our feet. Every day when I leave the house, I may walk over the place where a king killed a wolf in the Royal Forest of Stocket, one of the medieval hunting forests, where alder and birch, oak and hazel, willow, cherry and aspen grew. The living trees were cut down, their wood used to fuel the city's growth, its trade, its life. The ancient wood, preserved in peat, was found underneath the city. (The site of the killing is fairly well buried – the wolf and the king had their encounter some time around the early years of the eleventh century.) It's the same as in any other city, built up and over and round, ancient woodlands cut down, bogs drained, watercourses altered, a landscape rendered almost untraceable, vanished. Here, there's a history of 8,000 years of habitation, the evidence in excavated fish hooks and fish-bone reliquaries, in Bronze Age grave-goods of arrowheads and beakers, what's still under the surface, in revenants and ghosts of gardens, of doo'cots and orchards, of middens and piggeries, plague remains and witch-hunts, of Franciscans and Carmelites, their friaries buried, overtaken by time and stone. This is a stonemasons' city, a city of weavers and gardeners and shipwrights and where I walk, there was once was a Maison Dieu, a leper house; there were song schools and sewing schools, correction houses and tollbooths, hidden under layers of time, still there.

This city is a settlement where people have stopped, stayed, rooted down over what was here, over the foundations of another world, one we might call natural, layering up over time like the visible strata of an ancient tel, the components of which we can't, even through archaeology, be aware. When I was a student, I lived for a time near Tel Megiddo: Tell el-Mutesellim, the Hill of the Ruler, Chariot City, an ancient site in the Jezreel Valley in Israel where, over the span of 5,000 years, cities rose and fell, were built and destroyed and built again. Twenty cities were created there, one over the other; a city of Canaan, conquered by pharaohs, later ones built and destroyed by kingdoms which flourished, were defeated and flourished again, the cities of David and Solomon. They were places of war and conquest but also places where human beings lived everyday lives. On my journey home, I'd look from the bus window at the excavated site, a layered slope of ancient time, and measure what we know of them, of their inscriptions and their songs, their poetry and myths. I'd think of how much we leave behind, and know that we bequeath more than dust and stone.

March 9th

It's an icy morning when I follow more recent history and walk up to Rubislaw Den, to a place I haven't been to for years even though it's only a few minutes' walk from the house. Before the streets of the West End extended beyond the small centre of the city, 'The Den' was where Victorian students came with their professor on botanical rambles.

The professor was James William Helenus Trail, doctor, botanist, passionate collector of plants and insects. Born in 1851 in Orkney, Trail travelled and collected in the Brazilian Amazon before directing his attentions to the north-east, documenting the plant life of the area with meticulous care. For fifty years he collected specimens of weeds and introduced species, as well as keeping exhaustive records of everything that grew – or no longer did – in the city and beyond. The years following his appointment as Professor of Botany at Aberdeen in 1877 spanned the decades when the city was expanding, when the fine buildings of granite were being built, when streets were encroaching on unspoiled countryside. Trail recorded the destruction of plant life and the increasing restrictions on access to former collecting grounds: 'The extension of streets and buildings of course expels all plants alike from their former habitats.' During his classes, he liked to take students to explore the country roads and lanes leading to the old quarries near the edges of the city:

These last were then covered in part with trees ... with pools in some places and in others mounds of rubbish, bare on some slopes while on other parts, they were covered with tangled vegetation. Rubislaw Den and the old quarries were the scene of another of the excursions of the botanical class each summer and always gave a rich harvest.

I walk up through the long-established streets, parallel to the unseen Denburn flowing through its narrow wooded valley, now enclosed by the houses built by granite merchants for themselves. They made gardens to overlook the woodland of Rubislaw Den and in time, the place where Trail would bring his students to examine plants became inaccessible, sealed off, a secret enclave between two streets of houses. He wrote:

> The city continues to stretch out new streets rapidly to the west, and the plants that grew by hedges and fields, and on waste-places near farm-houses became more and more circumscribed, especially around Rubislaw. As the streets approached Rubislaw Den, the freedom of access to it had to be restricted, and it ceased to be visited by the botanical class.

(Trail also notes that on a map of 1746, the area to the west of this carries the inscription: 'Marshes and great stones, then chains of Mountains that stretch to ye Western Ocean', which is still a fair enough description of northern Scottish geography.)

Long gardens slope down from the houses to the road, their hedges and mature trees lining the low granite walls. Friends once lived in a house backing on to the Den, a family who, like us, kept unlikely assemblages of birds and beasts (and who perhaps influenced us by their wonderfully egalitarian and accepting attitude to their house-mates: a snapping turtle, a large parrot, cockatiels, a crow and a Labrador. They kept chickens for a time too, turning their garden into a scurry of rural sights and sounds). Our children played in the Den together but since they left many years ago, I've hardly been there at all. You can't get in except through someone else's back garden.

This time, I've asked people I know slightly if I might make my way through their garden and they've agreed. They're away, and so I have to pick up the back-gate key from the house. First, I have to breach high electric gates. There are bars and spikes and an intercom into which I say my name. The housekeeper lets me in and takes me through the hall. She is elderly and stooped and unlocks the French windows slowly before leading me onto a high terrace above the garden. She indicates where I'm to go before turning back to the house. I walk down the steps from the terrace to the grass. At the foot of the lawn, a small set of steps and a solid metal gate. It opens with difficulty. I tug on its hinges and squeeze though the narrow aperture. The winter must have rusted the iron. I am a Victorian child, opening a gate into a secret garden. I walk through and down a further small flight of steps into the emptiness of the Den. There's no one else around this morning. Scarce and now endangered hawfinches, stout-beaked birds and the largest of the finches, were once meant to have favoured the Den, nesting and breeding here. (They choose their places carefully, selectively. If there are any still in Scotland, they're said to be in the grounds of Scone Palace in Perthshire, or in the Royal Botanical Gardens in Edinburgh.) This morning, there are only a few wood pigeons clattering in the trees of beech and pine, hazel and sycamore growing on the slopes of this small valley. The water of the burn is clear and loud, filled by melted snow and rain. The crocuses are just beginning, the snowdrops still out in scattered patches of light and shadow. *Ranunculus ficaria*, the yellow celandine, is beginning to flower, the buckler ferns, *Dryopteris dilatata* and wild garlic growing under the birch and beech, the spruce and pine.

In the quick water, the pages of a dropped, drowned book turn and turn in the flow, as if there is an urgent message I must read, one which has already been washed away by the fall of water. I think of Professor Trail on a morning like this, bending, picking a specimen, showing it to his eager students. I take photographs and walk in the peaceful March morning, entirely alone, peering towards the backs of the houses. They're too far away to see, too high. People have built walls above the embankments, another manoeuvre to keep the world at bay. I walk to the end where the Den peters out under tangles of barbed wire coiled below the road. Above, I can hear the traffic of Anderson Drive. The day is continuing in the offices of oil companies on the other side of the dual carriageway, distant in some way other than place. I turn back, and it feels as if I'm not part of the busy world behind me.

When I see a couple of large Labradors approach along the path and hear the footsteps of their owner, I scamper towards the gate as if I'm an interloper. I squeeze myself through and back into the garden, where hellebores are just beginning to flower beneath the trees. I walk back up the garden to the terrace and knock timidly on the French doors. The housekeeper and I chat for a little, mainly about the birds we see in our own gardens, siskins and sparrows, different types of finch. She smiles as if she's not entirely used to smiling. She lets me out and I negotiate the slow, commanding pace of the electric gates and feel as if I have escaped or passed through an invisible corridor in time. It's only a short distance to my own house. My metal gate and fence, easily negotiable, are rusting, peeling, badly needing paint.

A few days later, I walk up beyond the houses to the dual carriage-way to where, from the other side, I had stood where the Den ends. From here, it's a steep drop to the level of the ground below, the coils of barbed wire there to deter curious passers-by. You could live here for a long time and never know it's there.

EARLY SPRING

March 14th

Since my walk on the beach, I've had my copy of *The Outermost House* near to hand. Today, I read something that intrigues me. I've read it before and have always meant to find out more. After a storm, Henry Beston found the body of an apparently endangered bird, one of the auk family, *Uria troile troile* – called in America a murre – on the beach and wrote that it was a creature 'whom men have almost erased from the list of living things',.and now I wonder about both the bird and the list. What, in 1927, was that list perceived to be? Even now, there's no list of living things (or no list of human computation); no list of what we have or what we had and have lost, or are losing. We don't know how many species there are on earth, what has already been erased in the processes, natural and unnatural, of time, what has lived for millions of years, evolved, become extinct before our ever knowing. There are guesses and estimates, and all are frequently revised and the numbers grow and shrink and change, as do the boundaries of what we seek. (We see most what we seek most, but the nature and ways of the world persist without us and our far from all-seeing eye.) How many species are there? What constitutes a species? Who decides? Who calculates the measure of our knowing?

When I begin to read about species numbers, astonishing facts fly at me; estimates of species numbers that vary between 30 million (an estimate now more or less discounted) and 5 to 8 million, only 3 per cent to 5 per cent of them vertebrates. There are one million identified insects and an estimated 30 million as yet unidentified – 300 lb of insect for every 1lb of human. There are, I discover, 500 species of aphid, 10,000 species of sponge, 100,000 species of mollusc. The most recent research suggests that there are 8.7 million eukaryotic species (species with a cell nucleus), of which almost 90 per cent of both land and marine species are as yet undiscovered. The University of Florida publishes a book of *Insect Records* which 'names insect champions and describes their achievements', their achievements including being the loudest, having the fastest wingbeats, being most tolerant of heat desiccation or having the 'most spectacular mating'. The smaller the entity, the less studied and known; in a paper published online, Professor Klaus Rohde writes: 'Most species of multicellular, unicellular and microorganism have not been described.'

Professor Rohde writes about the role of parasites and symbionts, the smallest of all creatures; of the fact that some are so specific to their host that if the latter faces extinction, so will the former. It is the minute, the marginal, he suggests, that have so far evaded our notice or interest. In 1986, a genus of cyanobacteria, the smallest of the photosynthetic organisms and perhaps the most numerous of life forms, *Prochlorococcus*, was discovered by a microbiologist at the Massachusetts Institute of Technology. In a paper published in *Yale Environment 360* magazine, Richard Conniff writes of the significance of the find that these minute and abundant bacteria produce almost 20 per cent of the oxygen we breathe. He writes too of the implications and effects of loss of species, of unintended consequences and the benefits of chance finds, of our sacrificing the habitats and lives of creatures and hence, without knowing, our own lives too, interdependent as we all are.

In a radio broadcast in 2008, Dr Eric Chivian, founder and director of the Center for Health and the Global Environment at Harvard Medical School (and winner of the Nobel Prize for his work with the organisation International Physicians for the Prevention of Nuclear War, discussing the book he co-edited with Dr Aaron Bernstein, *Sustaining Life: How Human Health Depends on Biodiversity*, talked of 400 bacterial microbial species found in the gut of a termite and suggested that a different termite might have a different 400. 'We know almost nothing about life on this planet,' he said, a thought that makes one stop and breathe in deeply.

As I'm reading about species numbers and the concept of 'species diversity', I notice that one name occurs again and again. It is the name 'Santa Rosalia'. I find it appearing in references, in the titles of papers, in 'roads to' and homages and revisitings. It's there in every form and often, and although every road seems to lead right there, where or who is Santa Rosalia?

In 1958, at an annual meeting of the American Society of Naturalists, George Evelyn Hutchinson, a professor of zoology from Yale, gave a paper called 'Homage to Santa Rosalia or Why Are There So Many Kinds of Animals?' The paper was based on ideas he derived from his observations of aquatic insects of the Corixidae family made during a visit to Sicily. With great modesty, erudition and charm,

Hutchinson describes the process of finding the place where he made his observations, a small artificial pool lying below the sanctuary of a church near the limestone cave on Monte Pellegrino where, in the sixteenth century, a stalactite-encrusted skeleton was discovered, together with a rosary and cross. The skeleton was that of St Rosalia, a twelfth-century descendant of Charlemagne who, on rejecting the world and the life of her wealthy family, lived in the cave until her death, being found only when, following her appearance in a vision, she was credited with having saved Palermo from the plague.

This paper, building on work presented the year before at a symposium at Cold Harbor in New York under the beguilingly throwaway title of 'Concluding Remarks', proved to be one of the most important papers in the study of ecosystems, one still regarded as seminal, not only as an observation of aquatic biology but as a landmark in the subsequent development of studies in ecology, every aspect of which Hutchinson had pioneered during his 30 years of work at Yale. A remarkable man of diverse and profound intellectual gifts, Hutchinson, who was from an academic background in Cambridge, was a limnologist, an expert in the study of inland waters, and an ecologist whose interests included biogeochemistry, radioecology, palaeoecology and the application of cybernetic theory to his fields of study. His mathematical model of the role of the ecological niche remains one of the important concepts of ecology. Regarded by colleagues and students alike as a figure of towering intellectual breadth and interest, he was reluctant to accept the notion of his being 'the father of ecology', crediting Charles Darwin with the title instead. In describing the pool on Monte Pellegrino, Hutchinson hints at his wide interest in art and history in noting that the bones of an extinct Pleistocene Equus – an ice-age forerunner of the horse – were found in nearby caves, and that a further cave holds Palaeolithic Gravettian-culture flint-burin engravings.

'There are at the present time,' Hutchinson wrote in 'Homage to Santa Rosalia', 'supposed to be ... about one million described species of animal. Of these about three-quarters are insects, of which a quite disproportionately large number are members of a single order, the Coleoptera,' and he quotes the well-known story he suggests may be apocryphal, of the biologist J. B. S. Haldane being asked by a group of theologians what might be known of the Creator from his studies, to which Haldane is said to have replied, 'An inordinate fondness for beetles.'

Santa Rosalia has expanded beyond her traditional role as patron saint of Palermo and is now honoured by the many writers of scientific papers on the subject who invoke her name as their patron saint.

I continue trying to find out what happened to *Uria troile troile* who, it turns out, wasn't endangered at all. It seems, in the course of taxonomic time, to have changed its name to *Uria aalge*, the common guillemot, those beautiful seabirds with head and back feathers of softest dark brown and gentle faces, the guillemots who shortly will be nesting in numbers, starring the cliffs south of the city on their rock-ledge nests.

March 20th

The day of the vernal equinox. It's Han's birthday. Both my children were born on equinoxes, Han the vernal, Bec the autumnal, and so their birthdays are always on the day of the equinox or the one before.

As usual, I check to see if there will be an aurora. I'm not surprised when I'm told there won't.

March 21st

The clocks change, move forward an hour. An hour's lost. It disappears into the arbitrary, unknown place where lost hours go. It will take me around six months to recover from this assault upon my own fixed, immovable, internal clocks, all set implacably to winter.

March 22nd

I'm engaged in an email discussion with a friend in Brazil about birds. We're comparing species and Linnaean names. He has lived in Europe and writes: 'You in the north have so little biodiversity ...' My first reaction is to be outraged. I'm about to go into cyber-battle in defence of our natural world but of course he's right. By comparison, we don't, and one reason at least (amid a complex range of other ones) is latitude. Knowing the truth of this removes my desire to argue.

In the past few years, I've become interested in the thought of latitude. By now, I'm a latitude fanatic. It's as if some special merit might accrue to the inhabitants of places with higher numbers, or northern ones at least. I'm secretly competitive. I check lists. I compare and am, for no reason that seems worthy even to me, pleased by the fact that I live in a place situated at a latitude of 57 degrees north, as if it promises or perhaps demands a measure of stoicism I do not have.

Of all things, it's latitude that creates us, forms us, makes us who and what we are. Like weather, it separates us; it makes us diverse, tree or bird or human, thins us out or crams us in, multiplying. Latitude is one of the vital factors in determining what's out there, who is out there, who lives where, shares our lives, our skies, our earth and thrives here, at 57 degrees of latitude, other than ourselves. I think about the colours of northern regions, of northern creatures, the grey, the pale, the dun and look around at the species I can see, the familiar, the ones you see on any day in most northern cities: birds, butterflies – in season – spiders, insects, and I note their colours, the gentle subtleties, their temperate, mellow beauty.

We don't have the brilliant, eye-piercing polychromatics of tropical species, the camouflage necessary to be part of a tree canopy, to blend with flower and leaf in sunlight. (It's their very beauty that has been to the cost of many tropical bird species; the astonishing variety of their colour that has damned parrots to centuries of trade and capture, and in many cases, extinction.)

Everything seems related, in one way or another, to latitude. The numbers of species, plant or animal, in any place is dictated by latitude, or more particularly, the conditions existing in specific latitudes: the nearer the tropics, that zone between the Tropics of Cancer in the north and Capricorn the south, 30 degrees north or south, more or less, the wider the diversity of species. (As with everything else, the number of species recorded seems to depend to some extent on who has studied what – vertebrates are the most studied; water creatures of all sorts, in both fresh and salt water, much less so.) But still, biodiversity is vastly more in the tropics. The reasons are complex; to do with area and geography, with evolutionary history and with the fact that higher solar radiation leads to faster rates of evolution.

Here, we just don't need such colour. We need to blend unnoticeably with stone and grass, or sea. Very little is brilliantly coloured from a distance. Up close, if you're lucky, you'll see unexpected colour on birds' wings, the glaze of green or blue gilded over the wings of starlings or magpies, but only if, for a moment's grace, they're near and bold and caught in sunlight. If you're lucky, you'll see the bright dab of red and yellow on a goldfinch's head. Insects too, caught in the right moment of light; inky, metallic, their tiny density of colour, their transparency of wing, the brief flash of the butterfly in summer.

In anticipation of spring, I hang up potential nesting places for robins and blue tits and great tits, wooden nesting boxes and small basket-like objects made from woven grasses and plant fibres in which I hope they'll find privacy and refuge. A particularly determined blue tit wrestles with the container of alpaca wool I've provided for nesting material. He tugs and tugs, and when he flies off with the small quantity he's extracted from between the tight wire squares of the peanut feeder I used to hold it, I go out with a crochet hook and ease out tufts of the thick, soft black wool to make it easier to remove. Now, the container hangs against the creeping hydrangea like a furred beast, waiting.

March 23rd

This morning as I work, prolonged, brisk, noisy, to-and-fro aerial combat is taking place across the rows of back gardens in shrill and shrieking sounds that rise and break and die away with distance. There's a flurry and flare of black and white, magpies and jackdaws in conflict over place or prey or just, knowing magpies as I do, for the sheer, annoying, glorious hell of it.

These raucous exchanges coincide with the first of the annual round of complaints against magpies to appear in the newspapers. (This is one of the few phenological events not subject to the vagaries of climate.) The person who writes every year to boast of the Larsen traps in her garden, the traps designed to catch magpies, which allow her to kill them by hitting their head against a wall, does so again. A man complains about the cruelty of magpies, using the opportunity to expound his personal theory of aggression. (I suspect him of being one of the people who read the Chinese military strategist Sun Tzu's work, mistaking themselves for Chinese generals of the Spring and Autumn Period.) The complaint against *Pica pica* is that by taking the eggs of small passerines, they reduce the population of 'songbirds'.

'Songbirds' are one of two suborders of the avian order Passeriformes, or 'perching birds'. The suborders are the Tyranni, the 'suboscines', and the Passeri, 'true songbirds'. Within the suborder Passeri, there is a further division into Corvida and Passerida. Magpies belong to the Corvida. Thus they are, by definition, songbirds and all reliable evidence suggests that magpies do not, in spite of what people think after having watched their activities, reduce populations of other passerine species. Cats do, pesticides do, removal of habitats and all the other phenomena, known and unknown which affect bird numbers do, but magpies do not. Magpies take the young of other species, not the adults. If they did, they would remove their future source of food. When they do take the young of other birds, it's for a limited time in the year unlike other destructive forces, which operate all year round. Magpies take the eggs and even the young of other birds because that is what they eat in order to live and I wonder, when I read the complaints, why magpies are considered less entitled to do so than other species. What they do may look cruel but this is the nature of life among birds and other creatures, even ones found in the purlieus of our garden. I'm suspicious of complaints against magpies. Many of them, I'm certain, are based on groundless superstitions or on the primitive idea that seeing one magpie will bring bad luck. I can talk only of my experience of having lived for many years with one solitary magpie. In the family, we all saw one magpie every day, often. From this potentially malign contact, none of the members of the household suffered any noticeable harm. From that single magpie we gained more joy, amusement, affection and interest than we could ever have imagined, as well as a lasting admiration and respect for the depth and scope of that one bird, and that particular species' astonishing, surpassing intelligence.

March 24th

We have such strange ideas of cruelty. A few days after I read the first of the anti-magpie letters, I'm walking in the network of tunnels extending under the wide main city-centre streets, under the high viaducts and Victorian bridges that give the city depth, demeanour and grace. I like these quiet under-streets, almost a parallel city below the streets of chain stores and traffic where shallow passages open on to the Green, a square of shops and restaurants where the thirteenth-century Carmelite friary once stood, where bleaching greens and markets, fish-wives and textile workers were, the ancient Mesolithic heart of the city. Steep steps lead up and away from them back to the teeming world above. On the path of the shortcut I'm taking one afternoon through one of the lanes, I see at my feet on the cobblestones a line of splashes, still wet in places but drying to darkness at the edges, carmine to garnet. It's blood, already drying but not long shed. It can only have been there since the recent rain, probably since last night. It's clearly human because there are none of the signs of its being anything else. There are no feathers, no fur, no animal detritus, no dank furred body, no sodden herring gull flattened in the gutter. Besides, nothing else leaves so much blood. Through the archways and passageways are nightclubs and pubs and places where the pleasures and conflicts of evening occur. This city isn't, I think, more violent than other cities. It's violent enough.

We all know of the violence of creatures, impelled by need for food or defence or conquest in the struggle for reproduction, violence which seems wanton but isn't. I've watched doves peck relentlessly at an ailing fellow, while I will remove it to isolation where I will attempt to feed it, revive it, ensuring that it will take longer to die, and I wonder every time where the mercy is and how I can conceive of or understand the true notion of it. I watch the hawk kill to eat, the spider which kills after mating. I think of those creatures who consume their young, possibly from the instinct to keep them safe from harm.

In wild creatures, violence is not purposeless. (We don't behave like 'wild animals'. If we behave badly, we behave as badly behaved humans.) There is always an explanation, even if we don't appreciate or understand it. What we find difficult is to be witness without judgement.

The American entomologist Jeffrey A. Lockwood has written extraordinarily of the violence at the heart of our lives, human and non-human alike. Explaining the work he did during a sabbatical in Australia on the family Gryllacrididae – an insect like a cricket – he writes about the ferocity of gryllacridids and the reasons for it, related to their size, the absence of the physical defence stratagems of many other insects such as poisons or stings, and he describes the extraordinary anger of gryllacridids facing danger, of how the desire to live makes the creatures so willing to kill.

In his essay, Lockwood writes too with admiration and respect of Dr Jeff LaFage, one of his teachers at university, who taught him a course in insect behaviour, emphasising to his students the evolutionary imperative of the gene, the necessity for scientific objectivity, for an avoidance of anthropomorphism. LaFage, a man Lockwood describes as 'a bioeconomic hardliner', taught him that pain perceived in others was not an experience that could be understood, being cognitive and not one of sensation.

It's in accidentally injuring one of his subject gryllacridids by trapping it in the cage door, that Lockwood realises his love for the creatures he's been studying. In witnessing with empathy the creature's possible pain, he feels that he is reacting not as an objective scientist, but as one who understands that 'as empathetic creatures we cannot stop ourselves from imagining the pain of other sentient beings', and when he writes of the murder, shortly after his return to the United States from Australia, of Dr LaFage during a mugging in New Orleans, Lockwood extends his empathy to what he describes as 'the angry, scared youth' who shot Dr LaFage, to the circumstances of a life which enabled him to carry out the act he did. In an elegaic conclusion, he writes of the anguish of loss, the terror of animals, the plight of soulless youths, a passage that, whenever I read it impresses and moves me as being extraordinary and remarkable for the depths of its compassion for both man and nature. 'Sometimes,' Lockwood writes, 'our own sadness is as much as we can bear.'

March 25th

As our hopes and expectations reach towards spring, the lightless gloom returns as a low, grey canopy of cloud and intermittent rain. The temperature rises. It's neither balmy nor even warm but the change is marked. One late afternoon, the rain stops and by evening, a pervasive and enveloping mist has infiltrated the city. It feels as if it has replaced the air. This isn't the haar, the mist that rises from the sea on the warm days of summer. The haar flits, moves in light gusts, in trails and billows. This kind descends and settles. Through this mist, the buildings on the other side of the street are barely visible, the lights in their windows dim lambent squares like the lights of a distant ship on a foggy ocean. At eight in the evening, as I'm standing on the front step, I hear the sound of geese although I can't see the sky. The sound is moving east towards the sea, growing fainter, goose voices melting into mist. As the sound diminishes, I imagine them flying across the city, out to the estuaries of the rivers, the Don and the Dee, where they open out into the North Sea. It seems too early for them to be leaving for their summer breeding grounds but perhaps it's not. They're the first of the season I've heard. Their direction of flight seems wrong. At this time of year, or a little later, they fly to Iceland but now, they're heading east. I wonder if the mist and rise in temperature is significant in their going. I put on my coat and walk out for a while into the darkness. The green of the traffic lights beyond my house hazes into the aura of thin electric blue given off by the floodlights lighting the office building on the corner.

Everything drips. There's hardly any traffic, just quiet and dripping. Alone in the mist, I pass no one and hear no more geese. I've never experienced this here before. It's different from the old fogs of Glasgow with their horrible opacity and subterranean fume of sulphur and coal. If this mist has any scent, it's salt. I've been reading Paolo Barbaro's *Venice Revealed* and his enumeration of the mists of that wateriest of cities: 'Once again, it's crept up on us slowly without anyone noticing it, this mixture of fog and night ... *Nebbia, nebietta, foschia, caligo ...* '

There are no footsteps, no voices, only the sound of water falling lightly from leaves. The sky is obscured; the moon is gone and the stars too. Every horizon is narrowed to a step in front, a step behind. Most of my own navigational aids have gone, some the same as those used by travelling birds – familiarity, sight, a sense of place – but for me, naturally, it's different. I have no choice but to stay earthbound, grounded and walking with boundaries, walls, hedges, pavements to guide me but for them, in these altered, distorted conditions, which is earth and which is sky?

In the air, we're little different from birds. We need the guidance of fixed markers. We lack their senses, their varied direction-finding manoeuvres but still, even with them birds get lost, crash into the ground as they mistake lights below for water, just as vagaries of horizon and visibility may cause planes to crash into the ground or the sea. In 1999, John Kennedy, piloting his Piper Saratoga towards Martha's Vineyard, crashed on a hazy night into the sea, killing himself, his wife and her sister.

'Other pilots flying similar routes on the night of the accident reported no visual horizon while flying over the sea because of the haze,' one crash report said. 'The National Transportation Safety Board determined that the probable cause of the accident was the pilot's failure to maintain control of the airplane during a descent over water at night which was a result of spatial disorientation.'

We have no automatic, fail-safe sense of where we are. Our view of ourselves in relation to place is regulated by the tiny parts of our inner ear. Humans and birds alike have otolith organs, utricle and saccule, and birds have a lagena too, but take away horizons and light and the sense of the ground below and we're all there together, floating, not flying, lost in space. How do we ever know where on earth we are?

A few days later, someone tells me that that same evening, they heard geese circling in the mist over Deeside. They sounded distressed, she said, their cries sharp and unfamiliar. She was sure that they were lost. A single goose was found wandering on the main Deeside road.

Orb-weavers and Wanderers

It's properly spring and I'm dusting again, carrying out the sometimes calming, sometimes panic-inducing activity that seems the only measured response to just one of the processes of nature. It has its dispiriting aspect because it will neither go away nor ever (a bit like leaf-collecting in autumn) be at an end. However assiduous I am, or anyone else for that matter, that fine distribution of the fabric of ourselves and of the physical world at large, the assorted particles, organic and inorganic, will fall daily, hourly, by minute and by second over everything; will accumulate in finely powdered layers, in a constant and never-ending continuum; a fall and drift that, were we to insist upon its total extirpation, could keep us all occupied for ever.

Spring sunlight seems to induce in most of us some level or another of action (or guilt and inaction) of the spring-cleaning variety. In me, it brings about a dusting, polishing, cleaning frenzy, a more fevered version of the winter's level, which though ameliorated, is thorough enough for winter. Spring is the moment when new, penetrating light unexpectedly strikes its first-morning glance through the filmy glass of winter, when domestic neglect in all its shabby, shaming annual collections is illuminated, and brightly. Dust, soot, ash, hazy coatings of

121

the winter's atmospheric accretions and – in my house – assortments of avian detritus: disintegrations of keratin casings from feathers, wisped feather particles, edible morsels hidden and forgotten and the fine dust some birds distribute widely from their wings, are everywhere.

Not only those, however – a rich and varied assortment of spider's webs (which, in the past three or four years, seem to have proliferated as never before) are revealed everywhere, tautly strung, glistening in the spring light outside windows, floating benignly above the warmth of radiators, forming thick, opaque sheets under furniture and behind pictures. They make pale and airy blankets visible over blinds, curtains, walls. Trailing wisps decorate the ornate cornices, the ceilings and the lights, trembling gently in the rising air. Corners are crossed by fine, drawn strands. In only a few are their mythic, to some, terrifying architects and creators still in residence. In many, homes have turned in the eternal processes of life and death into graveyards. Tiny, desiccated corpses hang like the aftermath of medieval executions. All winter, while we've all, in one way and another, been hanging in there, trying just to endure, spiders have been living, thriving, eating, dying in the space between the walls.

Once, the accumulated results of their efforts would have been swept away casually in the spring-cleaning surge but now, I hesitate. It seems somehow churlish to destroy other creatures' homes. It may be long habituation too but I'm accustomed to the demands and pleasures of communal living, and though I can't trace this back to one single moment of epiphany, for a long time I've accepted that the house isn't ours. This has nothing at all to do with the history of developments in progressive living (although I did once live in a commune, in the days when these things were modish, an experience that left me with fewer profound thoughts on the nature of our relationship to property and land than a lasting irritation with the domestic habits of others). When you've accepted that parts of your house may be yours in the technical sense only, that they're the sovereign territory of another realm and that your exclusive rights are subsumed by the needs and interests of others, it becomes easy, for, if you're going to live with other creatures, abjure cages and confinement, there's bound to come a moment when, like a state negotiating treaties, you accede suzerainty – in our case, over rooms and space rather than countries: the upper reaches of the kitchen, the utility room – and set about some sort of attempt at maintaining appropriate boundaries and relationships. From then on, the concept of 'giving ground' seems uncontentious, the only decent thing to do. But now, and increasingly, it's not only ground, it's ceilings too.

I've always had an amicable relationship with spiders, if 'relationship' is the correct word. I fully accept that it's one-sided, that their sole requirement is for me to stay as far away from them as possible, but I like and admire them as I'd admire anything described as 'small upon the earth but exceeding wise', anything of which it is said, 'The wisdom of the spider is greater than that of all the world together' because surely King Solomon and the accumulated wisdom of the many Akan peoples of West Africa can't both be wrong. I accept too that a number of the arachnid residents of this house may have dynasties of habituation dating from long before we ever arrived, and that time seems to grant as much of a grave and justified right to possession as the extended, legalised usury which underwrites most of the human habitation that takes place within the walls of our houses. Writing, or so we may assume, in Proverbs, King Solomon deemed three other creatures to be similarly endowed with wisdom – ants, locusts and rabbits – and although I haven't had sufficient dealings with the former two to know the truth of the assertion, formidable reputations precede both these amazingly successful insects. Of the latter one, I need no convincing.

You may, I think, live more easily in the world if you like spiders. Outside Antarctica, they're difficult to avoid, almost ubiquitous. Somewhere you're going to come across one or other of the 40,000 or so known species on earth. (Or perhaps even one of the possibly equal number as yet undiscovered. In Britain alone, there are 650 known species in 33 families.)

Spiders are, in many respects, a challenge – their wisdom notwithstanding. Of the many species living in close proximity to humans, they occupy, in the human mind at the very least, an ambiguous place in the hierarchy of orders of beings. Important in the chain of life and the systems of the earth as are all creatures, the unfortunate spider carries with it an overwhelming burden of human fear and dislike, as well as a distinct disadvantage when it comes to relative size. Often it seems that the only reaction to spiders – one regarded as the natural reaction – is revulsion, or fear, or panic, or all of these together, allied to the urgent desire to kill by any means.

It seems inexplicable sometimes that arachnophobia's so big among us, considering the paucity of spiders, in Britain at any rate, who present any significant danger to humans. (There are some dozen or so species in Britain which will bite but the likelihood of its happening is considerably less than that of being stung by a wasp.) But arachnophobia isn't rational, nor is it founded on weighing the risk and acting accordingly. Arachnophobia is real and painful, a response to a creature vastly different from ourselves. Our fears are unrelated to our size and power. We know perfectly well that, as with many other creatures for whom we feel fear, spiders have more to lose than we do, but there is no reason in fear. Its origins are subject to dispute – is arachnophobia a protective evolutionary response, innate and instinctive, or is it learned behaviour, a response found in societies detached and distanced from the natural world? Is it a response to dirt and to disorder, a possible inheritance from an association, however loose, with the events of the plague?

The way spiders move is often cited as another reason to fear them. Birds induce a similar panic, with fears of encroachment, entanglement, of stinging, piercing, biting, or worse than all those – danger unknown. Recent research at Rutgers University which tested fear responses in infants suggests that while the fear isn't innate, it is quickly learned, a useful mechanism in situations of potential danger. But in *The Private Life of Spiders*, Paul Hillyard suggests that not being afraid of spiders may be the learned behaviour, and that arachnophobes may not yet have learned to overcome their fear. Oddly, attitudes towards spiders in Britain, with its almost totally benign arachnid population, are more negative even than ones in countries where there are spiders that can do some degree of harm.

Hairiness is another attribute that does nothing for an unexpectedly hairy creature. (Hair, it seems, isn't much appreciated in any but the obvious cases.) The words 'great hairy spider' are not encouraging – in fact, add the word 'hairy' to anything not expected to have hair, and an element of horror and fear is introduced – spiders, caterpillars, even birds. Some years ago, in a newspaper report about ravens deemed to have been up to no good, they were described as having 'huge hairy beaks'. Had the report been true, it would have been an astonishing scientific discovery, the first historic reporting of the phenomenon. Hair where it's not supposed to be is useful for inspiring terror and disgust. The Regal Jumping Spider whose photo I see in a book, much enlarged, is described in the caption as an 'exceptionally hairy spider'. (It is, but in the photo it looks furry, woolly, beguiling, rather like the kind of child's soft toy that bounces off the end of a piece of elastic.) Hair is perhaps not much of an indicator of sinister intent but on a spider, it's useful, essential, a mechanoreceptor, allowing it to sense aspects of the environment upon which its life depends.

Spiders suffer greatly from their widespread portrayal as threatening and potentially dangerous, caught as they are in the endless human need and desire to be spooked by something. The film industry deals with spiders in much the same way as it does with corvids, using them to create and expand upon an entirely false notion of their propensities, roles and behaviour. A news story featuring research into arachnid behaviour will provoke a frenzy of hysterical headlines: 'BLACK WIDOW SPIDERS ON THE MARCH!' 'GIANT SPIDER NUMBERS WILL INCREASE WITH GLOBAL WARMING!' When research done at the University of Kansas suggested the possible move north of the potentially dangerous

American spider, the brown recluse (aka the violin spider, *Loxosceles reclusa*, found only in the Southern States), in reaction to global warming, the response was predictable, the most hysterical being from a news channel that assiduously denies the phenomenon of climate change. A spider expert from the University of California who provides arachnid information on his website writes wearily of the massive volume of incorrect information, the fevered tone of the reporting and general ignorance of spiders and their beneficial role and habits. Of the dangers of the brown recluse, he writes: 'Bites are rare. Ninety per cent are insignificant.' Another commentator says only: 'Brown recluses in New York? Bummer.'

Spiders are ancient. The first evidence of them dates from the Devonian period, 440 million years ago, in findings of spider ancestors suspended for aeons imprisoned in amber. Many fossil spiders look little different from any spider that might be seen today. By the Carboniferous period 330 million years ago, they were already highly developed. Evidence for their past presence is slight; as slight as they are, small, delicate, of a fabric easily lost or shrivelled or crushed, their bodies of chitin and cuticle, exoskeleton and carapace, turned in a moment to dust. Most of the ones we have were caught in amber, in a moment's dripping, encasing, hardening; time held, millions of years imprisoned within that enduring exudate of long-ago trees although some have been preserved too in fine lava dust which is light enough to protect the structures of their bodies. Nothing has preserved a web although a single spinneret, from which silk is woven, was found in Devonian shale in New York State. Doughty, small and remarkable as they are, spiders were among the species not only to have survived a minimum of two of the five great 'extinction events' that overtook earth, the Cenomanian–Turonian extinction event of 91 million years ago and the K-T, the Cretaceous–Tertiary of 65 million years ago, but to have increased their numbers afterwards.

I'm used to corvids, to the way they look and behave, to the fact that they have black feathers, pointy black toes, wings and beaks, and that their knees are actually their ankles. I accept that they are intelligent in ways and to extents I probably can't imagine, in addition to their many other qualities, but I accept that some of the human fear of birds I've encountered may lie in the fact that in physical terms, they're so unlike us: small and light and apt to do things we can't do, like fly. Birds are different enough but share at least some of the anatomy, and the vocabulary of anatomy we can recognise. Spiders, however – exceeding wise they may well be but they're also exceeding strange. Considering differences of morphology and scale, spiders are beyond even the otherness of birds. Spiders are, in their presence and their habit, in their life cycle, in their lives and deaths, almost unknowably alien and unfamiliar, not only in the words of their anatomy, in their strange and wonderful bodies too – in their exoskeleton, their cephalothorax and abdomen, in which all their organs are located, in their chelicerae, the jaws and fangs with which they deal so efficiently with living prey, the book lungs (the respiratory organs which look like book pages) and trachea with which they breathe, the haemolymph that circulates to all their organs, the web-making spinnerets and spiracles, the pedipalps, like jointed antennae on their heads, in males used for the transference of sperm. They live within an exoskeleton of chitin and cuticle from which they have to moult as they grow, increasing their haemolymph levels until the pressure cracks the old exoskeleton, allowing the spider to free itself, to grow, to harden the shell of a new one. They move on legs that end in thick scopular hairs like tiny feet, allowing them the exceptional freedom to walk up anything (although baths still seem to defeat those marching feet). They have, moreover, eight of these remarkable legs and between two and eight eyes – apart from the cave-dwelling species, who don't need eyes at all. No wonder spiders seem strange. We're two-eye-centric, suspicious of anyone, or anything, with more. We're vast, in spider terms, with dangerous hands, large and heavy feet, and have, in all too many circumstances, towards even the very smallest, homicidal inclinations. We do, however, have only two eyes.

It's often difficult to tell how many eyes spiders have without looking very closely, arrayed as they are about their head like a jewelled coronet or a set of fancy headlights. The necessity for good vision depends on species and habit – nocturnal web-catchers have the poorest, and the daylight hunters, the best. Some of the jumping spiders, the eight-eyed Salticidae, have eyesight of astonishing acuity. It's the Salticidae that are regarded as the most intelligent of spiders, their behaviour being described as more closely resembling birds or mammals than other arachnids. The arachnid genus *Portia* is deemed the most able of all, *Portia fimbriata* having shown itself capable of adopting and learning innovative strategic approaches to hunting, in techniques such as plucking the threads of adversaries' webs in particular rhythms in order to fool them, finding ways to overcome obstacles and in negotiating routes on the path to prey. In a piece of remarkable research published in the *International Journal of Comparative Psychology* in 2006, researchers from the University of Canterbury in New Zealand demonstrated problem-solving abilities in the species *Portia labiata*. The task involved escaping from an island in the middle of a tray of water, the results demonstrating that the spider was both able to learn, and to replicate learned behaviour.

Spiders, for all the fact that they can't fly, get about. They do so by 'ballooning', by producing their own means of transport, cleverly travelling on their own silk. Taking off from a high point, they thread out a line of silk to catch on the wind which carries the spider over distances, sometimes hundreds of miles, even out to sea, to heights, even to altitudes of over 6,000 metres. (Darwin recorded seeing small spiders on the rigging of the *Beagle*, 'far out to sea'.) Work done at Rothamsted Research showed that the silk thread upon which the spider travels is sufficiently flexible to follow the flow of air currents, allowing the spider to travel further, enabling spiders to reach and colonise previously spider-less places.

Spiders are everywhere in literature, too. They're there in nursery rhymes, in 'Little Miss Muffet', 'The Spider and the Fly' and in Archy's splendid poem, 'Pity the Poor Spider' in Don Marquis's *Archy and Mehitabel*. One of the best known literary spiders is the inimitable Charlotte in E. B. White's classic book *Charlotte's Web*, the story of a barn spider, *Araneus cavaticus*. (So affecting is the ending of the book that the author himself was overcome while giving a public reading when he came to the last words, 'No one was with her when she died.')

One of my favourite books on spiders, *Spiderland* by R. A. Ellis, was published in 1912. It is a book that employs a tone of mellow benevolence combined with a touching empathy for the subject, which is, while frankly anthropomorphic, also charming and wonderfully out of time. (The book, Ellis writes in the introduction, is particularly aimed at 'young folk'. Thinking about how 'young folk' today might respond to it suggests itself as a useful topic for a research paper on the exponential growth of cynicism in the past century.) Spiders are, Ellis writes with deep respect, 'great fighters and terrible cannibals'.

In spite of their intimate dealings with such formidable creatures, arachnologists in their naming of spiders (much as ornithologists in the naming of birds) display an admirably poetic nature, a glorious ability to convey character by description. Who could fail to be entranced by the Cloud-living, the Dew-drop or the Garden-ghost spider? Who could resist the colour and allure of the Flame-knee spider or the Fireleg, the Graysmoke, the Purple-bloom, the Red-bloom, the Filmydome, the Starburst or the Rose? Is there anyone who, on reading the names of the Robber Baron Cave meshweaver, the Hobo, the Government Canyon Bat Cave meshweaver, the Tucson recluse, doesn't hear the score of a spaghetti western, or see in front of them a cast of improbably glamorous stars chosen to act the parts of these charismatic loners?

Forty thousand species, habits, habitats and differing lives. Spiders are everything. They're dancers, strummers, drummers, stridulators;

they're hunters, trappers, hiders, lurers, lurkers, stalkers, watchers, preyed-upon and preying. Either solitary, conducting life more or less alone, or social and socialised, living in colonies, cooperatively, sharing labour, sharing food. (When it comes to eating, they're enzyme-injectors, liquefiers, ingesters of their prey.) Of their mating, when many males, much smaller and more vulnerable than their potential mates, actively have to seek out larger females, possibly over a wide area, Ellis says, with a degree of understatement: 'Lovemaking in the spider world is a delicate and dangerous undertaking,' and indeed it seems to be, males often failing to survive the encounter, although in *The Natural History of Spiders* Ken and Rod Preston-Mafham suggest: 'The risks run by male spiders during their reproductive endeavours have probably been greatly overstressed. Sexual relations are probably good most of the time, often excellent, and it is usually only males who push their luck at the last moment who run really serious risks of suffering premature death,' which is nice to know. (These brave fellows are given the name of 'kamikaze' males.) Kamikaze lovers, bringers of wedding gifts, practitioners of bridal bondage, in fact serial wrappers-up of things, of brides and egg sacs, presents, food gifts. Female spiders may be nest-makers or egg-abandoners or perfect, protective mothers, carrying their young, nurturing and cosseting and at the last, donating their bodies in an act that if it weren't undertaken by a spider, would seem weirdly sacrificial, to their offspring as food. It's their reputations as mothers that inspired the great French-born sculptor Louise Bourgeois to create her drawings and elegantly beautiful large bronze spiders, as odes to her own mother, a tapestry restorer, a weaver, like a spider. In an interview with a print workshop in New York, Bourgeois described her mother as 'deliberate, clever, patient, soothing, reasonable, dainty, subtle, indispensable, neat and useful as an *araignée*'.

Strangely, paradoxically, spiders, regarded by many as sinister or ugly, are producers of the world's most beautiful natural forms, apparently delicate in texture but of astonishing strength (a paradox too, like their reputation and their image, these tiny persisters, these refusers-to-give-up, are inspirers and encouragers for the faint-hearted everywhere). Their silk, a kind of liquid silk protein extruded from a spider's 'spinnerets', is stronger than steel or Kevlar and is the material used for constructing all webs, sheet webs and orb webs, tunnels and funnels and tubes and tangle webs, these complex homes and traps and hiding places.

I watch an orb-weaver, *Araneus diadematus*, spin a web across the corner of the back door in its careful measured process, back and forth as the lines of silk weave, extend, form the classic web, a standard of engineering perfection and exquisite beauty. It's impossible not to wonder, not to address yourself to this small and purposeful creature although you know that they come into this world knowing; to ask, what is it that makes you do what you do? How do you know it? On a day of powerful, gusting wind in winter, I watch a snow-covered orb web outside the sitting-room window shudder, balloon outwards, again and again, buffeted, but still holding.

Some orb-web spiders build into their webs stabilimenta – surface patterns and designs of noticeable, ultra-violet-reflecting white silk of stunning beauty, thicker than the rest of the web, often in the shape of a cross. It's uncertain why they do this, why they draw attention to their otherwise almost invisible, defensive, carefully placed webs. No one knows if it's to advertise their being there, to prevent birds from flying into their webs or to attract more prey. If it attracts more prey, it's likely to attract more predators too, and it seems as if we may never fully understand these fine balances in already precarious lives.

Under one of the recessed lights in the rat room, a delicate hammock of sheet web spans a corner, lit and silvered. I leave it until a corner pulls away from its mooring and folds, hangs, shivering in the air. Under the rusting frame of the slanting skylight in the lavatory at the back of the rat room, I leave layers of web so that I can watch the arachnid life of the lightwell, and the deaths too in the ghostly carapaces, the exuviae of cuticle and chitin spiders leave behind. The spider species are the ones I'd expect to be here, the most commonly occurring in northern climates, mainly of the genus *Tegenaria*, of the family Agelenidae, the sheet-web spiders, house spiders, *Tegenaria domestica*. (There are many species of *Tegenaria* including *Tegenaria gigantea*, which isn't all that gigantic.) We have garden spiders, *Araneus diadematus*, members of the money-spider family, the Linyphiidae, the ones commonly found in gardens, the ones to be seen scurrying, carrying their egg sacs.

By now, in the rest of the house, I know the territory of the behind-the-fridge spider, the study-bookcase spider, the back of the Orkney-chair spider, the stairwell spider and though their homes are visible, their inhabitants are slightly less so, but for all of us, visible and invisible presences, we need not impinge on one another's lives.

The day before Christmas, I watch a tiny orb-web spider making her way down the kitchen wall, the slow folding and unfolding of legs and her obliviousness to my presence feels like a welcome lesson about the nature of days and time. I remember an evening a few years ago, looking out of a window at the top of the vertiginous fourteenth-century Gothic tower of the Old Town bridge in Prague, a city that seems the essence of European life and culture. In the warmth of early autumn, in the fierce glare of the building's arc lights, a spider empire extended itself, an arachnid world of webs spanning stone lintels – thick curtains, mazes, veils of webs, all lit and brilliant. The empire swarmed, its active occupants busy with the insect life, attracted on a warm September evening by the dazzle. Below the castle on its hill, the lights of the bridges over the Vltava, both banks sparkling in the growing dark. Watching them, for a moment, I envied the spiders their lives, their elevated distance, the lofty height from which they could, in that place of history and war and change ignore, as they must have done for centuries, millennia, our follies and accomplishments.

They create their webs so quickly. Walking downstairs in the morning, I feel a touch across my forehead. I reach under a low chair and feel the softness of web across my wrist. There's a sheet web over the rearview mirrors on my car. The gutter that I cleared of leaves the previous day sparkles with a soft line of sheet webs. The pair of lovely origami rooks someone sent to me are draped one morning in silver threads as they stand on the study sill.

Like all the rest of us, spiders are being affected by changes in climate, by increasing rain and heat, by alterations in the timing of

135

flowering and growing, as dependent as the rest of us on every other species, plant or otherwise, around them. A study conducted at the University of Fribourg over many years demonstrates that climate change is affecting the times and patterns of spider migration and dispersal – the 'ballooning' processes that ensure their widespread distribution – and that these phenological alterations may 'reduce the resilience of spider species and increase their vulnerability to local extinction'.

As with every other creature, every species, family and order of being, we lose spiders at our peril. In a paper on the effects of land management on the abundance of spiders, researchers at King Juan Carlos University in Spain suggest that spiders 'do not enjoy an excessive level of public sympathy' before enumerating their many vital, irreplaceable contributions to the natural world as important predators in terrestrial ecosystems, as pest controllers in agriculture and forests, and useful indicators of fine changes in habitats and climate. Spiders in their turn are important food sources for birds, bats, amphibians and small mammals such as shrews. A recent steep decline in sparrow numbers may be attributable to a fall in numbers of insects and spiders. In *Silent Summer*, Richard Chadd and Brian Eversham survey the future of Britain's arachnid population, concluding that, in many places, spiders may be adversely affected by warming, rises in sea levels, flooding and drought, while Paul Hillyard suggests that one problem of spider conservation is that the data relating to the subject is both limited and uncollected as a complete body of evidence, and that in any case, too little funding is available for it.

There is, in this house, one alarming arachnid resident, a specimen of *Avicularia urticans* – a Peruvian pink-toed tarantula (although it gives no hint of pinkness anywhere) who sleeps the certain, unthreatening sleep of death in his display box fixed to the wall of the room called 'the sewing room' (although not much sewing tends to happen there). He has been resident now for many years, and while I might hesitate at the prospect of his live self, his dead one brings no fear, only manifold wonderment. Bought in London by Han, then a frank and advanced arachnophobe, as a present for her father during a school visit to the capital, *Avicularia urticans* travelled with her in his glass-topped case on the interminable night bus to Aberdeen, in what I can only assume must have been an orgy of self-inflicted horror. Looking at him, at his leg span, size of abdomen and cephalothorax, I realise and accept that I can be as arachnophilic as I choose, as loftily moral and northern, as safe and unthreatened as I might be in the absence of any spider here with untoward habits of hiding under lavatory seats, or in beds or any of the other day-to-day manoeuvres carried out by more southern types, safe in the knowledge that it's very unlikely I'm going to encounter anything the size of this one.

Arachnophobia is clearly a condition from which one can recover. Not long ago, Han and I were looking through the photos on my phone. As I flicked through them, she spotted one I took recently of the spider I found resting peaceably on my kitchen floor. 'Hey,' she said, 'you've got a photo of my spider!'

No, I pointed out, it wasn't her spider. It was my spider. The fact is, my spider closely resembles her spider, the one she photographed recently resting peaceably on her kitchen floor in London. It's not really surprising that you can't tell them apart. They're both *Tegenaria domestica.*

I read accounts of people and their relationships with spiders. In her book, *Good Observers of Nature: American Women and the Scientific Study of the Natural World 1820–1885*, Tina Gianquitto describes the work of Mary Treat, a naturalist who lived and worked in New Jersey during the middle years of the nineteenth century. In one of her essays in the journal *Home Studies*, Treat writes of keeping spiders and their nests in large glass sweet jars for purposes of observation and describes how she puts moss and small flowering plants into the jars too, so that visiting ladies might be able to admire the plants without realising that what they were looking at were the homes of large spiders.

In *What the Stones Remember*, Patrick Lane writes, in addition to wonderful observations of spider behaviour: 'I measure friendship by those who are the friends of spiders and those who are not,' making me realise that until recently, I harboured the notion – now revealed to me as false – that most people don't like spiders. Now, I ask friends, acquaintances, anyone I meet, about their feelings towards spiders and more than I anticipate say that they like them. One or two profess a kind of love. Some tell me about their 'pet' spider, or at least the one they observe daily although, as Han and I demonstrated, making fine distinctions and recognising individuals can present certain difficulties. One friend says that she's noticed that even when she sweeps away a spider's web, another one replaces it swiftly, in exactly the same place, which I've observed too. She removes errant spiders with a glass and a

Blondie CD she keeps handy for the purpose. People have their favourite ways, I discover, of spider removal or relocation. Another friend prefers a postcard of I. M. Pei's Louvre Pyramid at dusk. I use a bookmark of a Leonardo da Vinci drawing of the hand of God from an exhibition at the art gallery.

It's as I'm trying to avoid damaging webs, peering to see if a spider or spiderling is in the way of the nozzle of the vacuum cleaner or the feather duster that I think of the Jains, adherents of a religion that offers protection to the smallest of living creatures, concerned with the soul and its ultimate liberation from the samsara, the cycle of birth and rebirth. Every living being has a soul trapped in samsara, a soul that will attain bliss only through the ending of a cycle. Non-violence and the avoidance of killing are the fundamentals of their belief, their vegetarianism scrupulous and thorough. No potatoes or root vegetables may be eaten because earth-dwelling creatures may be harmed in the harvesting of them. Small muslin bags are used to filter water to trap waterborne insects. Ascetic Jain communities sweep the ground before them to avoid stepping on small life forms. Travel during the rainy season is not undertaken, the risk of harming creatures being too great, while cooking after dark is prohibited because fires attract insects to their deaths. One particularly observant group wear cloth mouth-shields to avoid ingesting tiny airborne insects, and on inadvertently killing one, a formal expression of repentance is made: 'I ask pardon from all living creatures. May all creatures pardon me. May I show friendship to all creatures and enmity towards none.'

I reflect on the demands put upon their lives by their vigilance and again, I'm awed before such rigour, such care. I can emulate it only to a limited degree but when I find a spider wandering, in danger, I pick it up, or urge it to safety on my sleeve or onto the bookmark or a handy piece of paper. *Where do you want to go?* But I can't ask it, and so it's at my whim that the spider is carried to the door, to a window, set down at random at a place of my choosing. To lessen the dangers (and my own subsequent feelings of guilt), I buy the resident spiders a little vacuum, not for their use but for mine, a long clear tube with a blue lid in which there is a motor and a mechanism that gently sucks the spider into the tube from where, the theory is, it can be easily released elsewhere. On the few occasions I try it, the spider doesn't appear to suffer any harm but just operating the thing makes me realise that being sucked into a tube must be, at best, a disorientating and probably not entirely welcome experience, and so, except for rare and difficult situations, I revert to the old methods and use the hand of God instead.

SPRING

April 5th

A very small great tit is flinging himself repeatedly against the glass of
my study window. For days now he's been here as I work, hovering,
clinging momentarily to the ivy; swinging, launching himself towards
the window and then *click*, as the cartilage of his beak hits the glass.
The weight behind him is slight; only that of his feathers, his scant flesh
and hollow bones. I watch him all one afternoon as I work outside at
the table in the garden in the unexpected sunshine. He leaps up fluttering, a tiny blur of wings against the smooth, unknown darkness of
glass in which he sees, probably, only his own reflection. It's spring, and
his nest must be nearby and as a territorial bird, he wishes to defend it
from the enemy in the window. He attacks repeatedly, and after a bout
of valiant flinging, flies off into silence. In minutes he's back, swinging
on a thin strand of ivy, wings whirring almost to invisibility like a hum-
mingbird's, only their speed keeping him aloft. *Click, click.*

Chicken begins to nest. The annual progress of the pulling of light
cables, the tearing of paper, the shouting and displaying is in full and
vigorous progress. The house is loud in the early morning with the
sound of plastic food dishes being clattered around floors. I'm not sure
if it's just me, but it seems that the winter has sapped us all. At first,
even Chicken seems half-hearted in her enterprise although as usual
the newspaper nest is established under the kitchen table. The nest

grows and is abandoned for a day or so, and is begun again. She lays a first egg.

The weather has become unexpectedly fine. With the sun, the hydrangea on the wall begins to grow. It grows so fast that soon it covers the alpaca-wool container and the nest boxes and the nesting pockets, and when I see the tits rushing in and out of the foliage, I don't know if they've nested inside their custom-built houses or have constructed their own quarters and ignored mine.

April 7th

Spring in a supermarket car park. Two tiny flickers of feather fly towards me at speed as I walk to my car. They're super-quick and flighty, all sexual impulse and sauce and energy. They whisk past me, very close.

'Near miss there!' someone passing says. All afternoon, I imagine scraping small birds from my eye.

Again and again over the days of nesting and breeding the great tit is here, all day, throwing himself against the window. In an oddly snooty statement about window-pecking, an American bird organisation suggests that the behaviour is explained by the fact that birds have 'neither the ability to reason, nor the capacity to understand the concept of reflection'. It was probably written years ago and hasn't been updated because it doesn't seem to take account of the change over recent years in knowledge about birds' highly developed brains and cognitive skills. In a paper published in 2005 ('Cognitive ornithology: the evolution of avian intelligence'), Professor Nathan Emery of the Department of Animal Behaviour at Cambridge wrote, under the heading 'Theory of Mind': 'Five years ago, the idea that a bird could think about another's mental states was preposterous.' Advances in neuroscience and avian neuroanatomy have made it possible to gauge more accurately the nature and capacity of birds' brains, while studies in cognition and social behaviour demonstrate the previously unimaginable complexities of the behaviour and relationships of other species. It's not just the larger, more famously 'clever' birds such as corvids and psittacines who demonstrate sophisticated levels of behaviour and understanding but also birds such as chickens, who have been shown to understand causal relationships, or starlings who can identify atypical patterns in sequences of sound. It's true, tits don't appear to understand the concept of reflection but very few species on earth do, or very few of those tested. The only creatures so far to have demonstrated the ability to recognise that they're looking at their own reflection in a mirror are humans, elephants, a few of the higher primates and magpies. I don't know if this bird can reason but whether he can or not, his effort seems vast and I wonder what I can do to help him. Hanging balloons or old CDs may help deter the behaviour. Covering the windows with paper is another suggested method although the paper may be torn away by the bird who thinks the adversary is hiding behind it. (If this is the case, is the bird demonstrating 'object permanence', the ability to understand that objects remain even if one can't see them? Is the same true for an imaginary foe?) If I did

145

cover the windows, I'd have to cover all the study windows because his enemy would be in every one of them. And if I did, he might try other windows, as studies have shown some birds do. I could cover all the windows and plunge the whole house into temporary springtime twilight but it would take some effort (and scaffolding) to cover three storeys of windows, back and front, and besides, the bird's persistence shows that superficially at least, he's unhurt by the activity. He's fulfilling a role, and even if it has been distorted by the human invention of glass, it appears more distraction than harm. Still, I regret what seems to me the effort of his flinging. I close the curtain over where he is but of course it doesn't diminish the reflection. *Click, click. Click, click.*

We don't help birds' unfortunate relationship with glass by our historic habit of building cities, places full of glass and light, in the middle of their migration paths. In her book *Wild Nights* Anne Matthews, writing about the wildlife of New York, describes the activities of birdrescuers who search for injured birds among the multitudes using New York as destination or flight path (Central Park itself is an astonishing centre for migrating birds, situated conveniently as it is on a northern migrating route.) She writes of the ones who collide with high windows, become lost, or experience one of the other mishaps that may befall a bird in a tall, lit city. One rescuer, Rebekah Creshkoff, describes finding a scarlet tanager (a smallish bird of the cardinal family) in a revolving door. Putting it in her pocket, she sets off on the subway to Central Park, releasing the bird at the 59th Street entrance. The bird flies off immediately. On being asked if the tanager had seemed grateful, Creshkoff replies, 'Do any of them? Sometimes I think so. It's like kids; they never call, they never write.'

In Toronto, another unfortunately placed, high-rise city, a splendid organisation called FLAP, the Fatal Light Awareness Program, urges office owners and householders to switch off night-time lights to reduce the numbers from among the 50 million migrating birds flying into windows. They too collect injured and dead birds, trying to rehabilitate the former and carefully use the corpses of the latter for the purposes of teaching, research and gallery display and offer useful advice on helping prevent birds flying into windows.

Birds, as all creatures in cities, have to adapt. Some will be urban by choice, others by circumstance as we have moved in and taken over their habitats. In addition to glass, urban birds face a variety of other challenges including noise, light and atmospheric pollution (to say nothing of our presence). They have to find mates, defend their territory and if they do it with the aid of their voice, they have to make themselves heard above the noise of cities. Male great tits in cities sing different songs from rural ones. They sing higher, faster songs, songs that can be picked up through the strands of urban noise. (It's known that birds develop 'dialects', particularly over large areas; that the calls of birds from different areas display differing, variable features. Another study into song patterns and adaptations in urban birds suggests that they have difficulty communicating with members of their own species from the countryside.) Urban birds may have bigger brains than their rural counterparts – in a recent paper, 'Brains and the city: big-brained passerine birds succeed in urban environments', Alexei Maklakov and others in a study at Uppsala University link passerine brain size with the ability to live successfully in urban environments. Light pollution affects birds, altering the timing of their singing, their egg-laying and breeding behaviour. (In cities, we must all have had the experience of hearing birdsong in the street in evening darkness.)

Living closely with humans, some birds learn to distinguish between individual humans, usually when that human is seen to pose a threat. Crows can single out people who have appeared to pose a threat from among large crowds, and although the behaviour was thought to be confined to corvids, research at Duke University in North Carolina has shown that mockingbirds can do the same.

This bird outside my window is an 'urban adaptor', a creature that has learned to live with us. In the field of urban ecology, the terms used of creatures are 'urban exploiters', 'urban adapters' and 'urban avoiders'. The 'urban exploiters' are the omnivores, the hardy, permanent residents, the rats, pigeons, sparrows, the ones who live on what we leave. The 'urban adapters' are the species such as corvids, the hardier of the small passerines, the tits, robins, wrens, opportunists who can adapt to change, find the ways to learn to live in cities, the ones who will, with luck, thrive and populate our gardens. Both are 'synanthropic' creatures who benefit from opportunities to feed and shelter provided by an urban environment. The urban avoiders are all the ones who can't adapt, who suffer if their habitats are overwhelmed in the onward rush of expansion and building, who are most liable to harm or to extinction.

Writing in their article 'Urbanisation and Development' published

148

in the book *Silent Summer* of the myriad factors – temperature, traffic, building, pollution and more – affecting urban wildlife, Kevin J. Gaston and Karl L. Evans say, dispiritingly enough: 'Determining how these various factors have interacted to influence urban wildlife is not straightforward as wildlife monitoring schemes in the UK have until very recently ignored urban areas almost entirely.'

Looking at this small bird, I wonder how long he has been here and what has affected his life since the great avian expansion of the Cretaceous period, how he fared during the ferocious egg-collecting and feather-wearing frenzy of the nineteenth century, the pesticide storms of the post-war years and now, the factors affecting his life about which we don't even know.

It's not only in the lives of creatures that there are divisions between city and country. In his article 'Resistance to Urban Nature' published in the *Michigan Quarterly Review* in 2001, the American environmental author John Tallmadge writes of the ideas that have led not only to the view that nature, or nature in its wilder forms, is not to be found in cities, but of their influence on views of the relationships between man and nature. There's a perception that all virtue is invested somewhere beyond, somewhere free from the artifices and influences of man. Quoting the American cultural historian Leo Marx, who describes these ideas as 'pastoral idealism', Tallmadge suggests that they were ideas with their origins in Greek poetry and taken up by writes, artists and 'idle aristocrats' over centuries and he decries them for their effect of allowing us to 'love the green world' while we continue to destroy and build over it in the furtherance of what we designate progress. By idealising the rural, the wild, the empty, he suggests, we are given the freedom to undervalue and to deal with the urban as we please.

In examining the concept of 'wilderness', the distinguished American historian William Cronon reinforces these ideas, contrasting concepts of wilderness as they developed from the biblical to the Romantic, from the perception of it as a place of raw danger to later ones of wilderness as a sanctuary where one might find exaltation of the soul, and still later, as natural a counterpoint in its purity and virtue to the degradation of civilisation.

We live in a small country, in a crowded continent, one where human influence can be felt fairly heavily in most places. (Tallmadge also writes that his backyard is wilder than most of Europe. Not having visited his backyard, I'm not in a position to judge.) There's little wilderness – not in a small island – even in the northern reaches of Scotland, not in the high Cairngorms or in the far northern reaches of the country because everything is owned, managed, controlled, reachable from any of the cities within a relatively short time. A project to protect an area of rare habitats in the north-west of Scotland, one of wildcats and golden eagles, freshwater pearl mussels, rare Atlantic oak, birch, hazel and peatlands, involves at least six separate organisations working in conjunction with local landowners: estimable and valuable and much to be welcomed as the project is, it suspends in linguistic or logical space the idea of 'wild'.

The ideas are familiar, the belief that the distance we have come from a state of 'true' nature is a measure of the corrupting hand of man, and the belief that only in places far from people lie the paths leading to our return to a state of prelapsarian grace. We decry the effects of society and humanity upon what we believe to have been unspoiled, wanting it to remain or return to being unspoiled in order that we ourselves should be able to go there to restore ourselves, to recover from the turpitude of the cities in which we live. We seem to want wilderness for ourselves alone, for the possibilities of finding ourselves, or being at one with ourselves, or for a degree of perceived spiritual benefit that is to be gained from the freedom to escape from the world as we find it, an escape from time itself. This isn't to suggest that the hand of man is not and has not been infinitely corrupting and damaging but the idea in itself seems to suggest that it is humankind, driven by a set of innate impulses, that is responsible rather than the complex structures of politics, society and production that underlie what, in yearning for wilderness, we most decry. It's a view that suggests that one set of moralities and inclinations is represented by what is urban while the rural, the 'wild', the non-urban, represents entirely another and in achieving greater distance between oneself and a city, every step is a stopping point on the path to some greater moral certainty or superiority.

Most Scots now are urban, for the same reasons that a majority of the earth's people now are; from historic necessity, poverty, ambition, opportunity. Here, we do urban blight as thoroughly as many other places. We might even be considered as experts in the field, the very name of Glasgow having been for many years a watchword for everything undesirable in urban life, although even from the centre of that vital, fascinating city, you can see the hills. In Scotland, countryside is near. It never was a distant, idealised place to which one could only aspire, an aspiration limited by money or by time. While we may still share the universal tendency towards nostalgia that often dominates our relationship to place, the gazing backwards to a past that may or may not have existed, immutable as the past is, free from the threats and dangers of the present, Scotland's nostalgia lacks the idyllic vision that underlies much pastoral nostalgia, the belief that it was better not only back then, but back there. In Scotland there's a clear-eyed suspicion (if not an actual certainty) that that belief neither is nor was true.

Our landscape shapes the nature of our sentiment. We have few of the pretty, striking villages found in places further south, in southern England, France or Italy. Our climate is sombre, our mood, our stone, our mode of building against the weather. Our villages tend towards the linear, the low and the stark. There are country towns and villages of great and solemn beauty but of little apparent comfort. Scotland was poor, and it may be the memory of poverty, of unyielding land and harsh climate, that prevents any untoward yearnings. History in Scotland is remembered with degrees of pain and remorse, as much for the perceived failures of the past as for the defeats, reversals, dispossessions. Concentration on the land clearances of the eighteenth and nineteenth centuries is often criticised as being a focus for unwarranted historic introspection but for me seems a connection that lies beyond a narrow vision of the history of a single place, one that resonates with universal experiences of mankind, a reminder of the evanescence of the moment with its fissile certainties, and of the apparently unchanging nature of the dealings and interactions of the human world.

Our reading too shapes our sensibility. For most Scots of my generation, education in poetry (as in all literature) was founded on English rather than Scottish texts, on Wordsworth, Coleridge and Keats, Byron and Shelley and it was only later that many of us discovered the poetry of place through Sorley MacLean's vision, with the beauty of its memories and ghosts:

> Time, the deer, is in the wood of Hallaig . . .
> From the Burn of Fearns to the raised beach
> That is clear in the mystery of the hills
> There is only the congregation of the girls
> Keeping up the endless walk, coming back to Hallaig in
>   the evening.

Or Norman MacCaig's:

> And the light floods down
> revealing mountain and flowers
> and so many shadows. If only
> a merlin would hurtle past, that atom
> of speed, that molecule of light.

While judgements about the superiority of country over town may not be exactly those of other places, when judgements are made, they're even more marked in Scotland. They don't stem from any spirit of romanticism, but from the seemingly eternal moral requirement for there to be at least some suffering and discomfort behind every enterprise. It's an ethic that pervades all Scottish endeavour, and ensures that the dangers of experiencing too much pleasure are sensibly avoided. Some activities are deemed worthy and moral and some are not. They might all involve the same expenditure of energy, the same distance walked, the same investment of time, but to walk in the hills is worthy, to walk in town isn't. The worse the weather, the more morally sound the undertaking. Here, we're still strict observers of the philosophy of *pain is good*.

April 8th

A jackdaw pauses once or twice on the rat-room roof, the first this year. A beautiful bird, head feathers magnificently black and startlingly grey, eyes sharp and silver. I know why he's here. I see jackdaws in the garden only in spring and only for a short time even though they nest in trees nearby, and I'm used to hearing their admirably argumentative voices calling and chatting from roof to roof. The supra-efficient monitoring service that keeps corvids vigilant has told it that a blackbird is nesting in the ivy. My first reaction is to think of defences: barriers, noise – what? Shall I take up residence in the garden? Shall I wield my anti-sparrowhawk red plastic trident, the one with which I defend doves from raptors? Before I accept the idiocy of my reaction, the unthinking foolishness, the inconsistency of this particular position, the wrong-headed, interventionist sentiment that usually I deride and try so arrogantly to correct in others, I reflect on the nature of my feelings, the desire, unthinkingly, to protect that which no one can protect. I reflect on disparities of size, on the ways in which we institutionalise our ignorance, raise our prejudice to the level of campaigning right-eousness. What are 'songbirds'? Why are robins considered something they are not? Why do we think one type of creature is more deserving of food, love, protection, life, than another? I value jackdaws as much as blue tits or great tits or sparrows or anything else that is nesting in my garden.

April 9th

A friend from Washington State, a fellow corvid enthusiast, emails to tell me about the newly fledged offspring of her two known and loved garden crows, Roarke and Giselle. Their fledgling is on the ground. She is anxious about the neighbour's cat. She wonders if she should take a day off work to keep guard. I think of them, of us, all over the world, everywhere, all of us anxious, powerless people.

There's a flicker in the grass of the front lawn. I pick up a struggling, beating knot of feathers; a greenfinch, feet and legs tangled in fine black thread. I carry him inside, hold him in one hand and with the other, with tiny sewing scissors, snip slowly until he is free. He is small, fragile and patient. On examining them, I see that his legs look undamaged.

We therefore have no further business one with the other. I carry him back to the garden, stand him firmly on my open palm and offer him the air.

April 10th

A friend who is out for a walk phones from somewhere in Deeside. He has spotted a small bird he thinks has been abandoned.

Just as Wi-Fi is a special blessing to those who keep small and chewing pets, every spring I discover that in deterring the enthusiastic bird-rescuer, mobile phones are invaluable. Now, every nesting season when someone phones me to say that they're out for a walk and that there's an infant bird of unidentified species *on the ground* and may they pick it up and bring it to me, I'm able to suggest that they leave it in peace to the ministrations of its parents, that they continue their walk, turn off the phone. Had mobiles been as common or affordable twenty or so years ago, I probably wouldn't have had any dealings at all with birds. As it is, with accumulated wisdom, or whatever the amalgam of experience and good sense might be, I have abjured bird-rescue for ever. It's not that I wouldn't willingly fill the house, as I did once, with orphaned, lost, abandoned birds, it's now possible to make the decision in the light of facts. Greater knowledge encourages me to caution. (It's not that I don't want to look after, to nurture and form relationships with other creatures, not that I don't value the ones from whose lives and presence I have derived the greatest impetus to knowledge and to joy, it's just that now I know that a small bird on the ground is more likely to be learning to fly than it is to have been abandoned. Any other circumstance, I'll consider. By this general rule, I'd still have taken in magpie, starling and crow.)

I ask him a few questions about the bird, where it is, how it looks, and clearly, it's not in need of rescue. Leave it alone, I say, walk on, it'll be fine.

The days expand into a week and the bird's still here, clicking. Small lines of white spatter the window. He sometimes stops on his ivy and nibbles on invisible insects. He peers towards the window and begins his ineffectual combat again.

I go into the garden very early. *Click, click.* As I work at my desk later in the day, he's moved to the middle window, briefly, but comes back again to the one beside my desk. It's the middle of April, the weather warmer than we'd expect. Chicken is still nesting. I wonder at the patience, the persistence of the behaviour, the tedium of sitting on eggs.

There are mornings when I think the bird has gone and I wonder if he's given up, or has finished nesting, or has met with one of the fates small birds frequently do meet with, but by mid-morning, he's back and clicking.

April 15th

It is almost a year since the Deepwater Horizon drilling rig exploded in the Gulf of Mexico, inundating coastal Louisiana, Mississippi and

Alabama with oil. The topic was urgent here. Everyone talked about it, knew someone who had some relationship to the disaster, someone who did work for one of the companies involved or didn't, someone who worried, some forewarned soothsayer who knew that one day it would happen, or someone who believed it never could.

Oil's everywhere here. There's no forgetting and no escape. Oil's on every sign, on car number plates, in front of every building in the business district of banks and stockbrokers and company headquarters. It's in the heavy granite name signs, grandiose and imposing like the tombs of kings, memorials to who knows what. The names and the things they offer or promise hint at the unknowable and strangely fascinating: offshore and undersea, wells and drilling and pumping, explosives and fluid, mud and tools and oilfields. They use words and names that suggest vastness of scale or heroism of undertaking, the names of the many Greek gods of the sea: Poseidon and Triton, Thetis and Tyche, fields and flows and pipelines, all this to coax or to wrestle that strange dark substance from the foundations of the ocean.

It was only days after the disaster last year when new neighbours moved into the house next door. Owned by an American oil company, it has been inhabited over the years since we came here by a succession of company employees who arrive, stay for a couple of years, do whatever mysterious things are done in the pursuit of oil and then move on. Before I'd had a chance to go round to welcome them, we met one morning as we were carrying shopping in from our cars. We stopped for a few moments to introduce ourselves, shaking hands, standing in the sunlit street for a while to talk. I asked, as one does, where they came from. 'Louisiana', they said and instantly, the morning felt different, everything that might be said almost igniting, unsaid in the air between us. When, after a breath of hesitation, we did speak of the oil spill, it was almost casually, about the possible effects, the impact on wildlife and on Louisiana's state bird, the brown pelican.

'Bad timing,' my neighbour said, 'it's breeding season.'

A few months later, an American nature magazine I read had a cover so beautiful that it looked at first glance like a carefully posed painting. It was executed in Rembrandt colours, sombre and glowing deep brown, ochre and gold but it was a photograph taken at a rescue centre in Fort Jackson, Louisiana by the Spanish photographer Daniel Beltrá of a group of eight oiled pelicans. They stand with an air of unending, stoic patience on the folds of a stained white sheet placed there to soak up the oil being washed from their feathers.

I remember it again as I read: 'A new danger, moreover, now threatens the birds at sea. An irreducible residue of crude oil ... remains in stills after oil distillation, and this is pumped into southbound tankers and emptied far offshore. This wretched pollution floats over large areas, and the birds alight in it and get it on their feathers. They inevitably die ... I am glad to be able to write that the situation is better than it was. Five years ago, the shores of Monomoy peninsula were strewn with hundreds, even thousands, of dead sea fowl, for the tankers pumped out slop as they were passing the shoals – into the very waters, indeed, on which the birds have lived since time began! ... but let us hope that all such pollution will presently end.' Henry Beston wrote in *The Outermost House* in 1926.

The week is turning into a fortnight. Every day I hear the sound of the bird at the window as I come downstairs in the early morning. *Click click. Click click.*

This afternoon, I watch an agitated oystercatcher running on the wall of the formal garden at Drum Castle, calling, piping, willing me and the few other people there to go away. I've come to wander round the early spring planting, the beginnings of the garden that'll be exquisite, in summer the walls heavy with roses and filled with the scent of box. Eventually, the other people drift away and I lurk a while behind a hedge until the bird stops calling. Within moments, its two chicks, grey, fuzzed, long-legged, are being urged into the flowerbed below the wall.

LATE SPRING

April 18th

Passover arrives, and the family comes home to celebrate. We conduct the Seder, that extended ritual of memory and feasting, with the usual mixture of enthusiasm and chaos, the latter encouraged by the fact that we all have different versions of the required text, the Haggadah. (The ones in my collection have been picked up wherever the Haggadah is to be found. It's serendipitous, and includes one saved from a flood in New York; a couple given away by the company which makes Matzot, the unleavened bread we eat for the occasion; the edition I had as a child with moving tabs and flaps with which one can, on a whim, drown pharaonic armies or find infants adrift in baskets among thickets of bulrushes; and two recent ones with arty illustrations and innovative typography.) As a result, the ceremonials are interrupted constantly by a refrain of, 'Where are we?' There are none of the seamless Hebrew recitations of my childhood undertaken by a past, cheder-educated generation of father and uncles. This time, it's me or nothing.

We set the table with the ritual items, the parsley and salt water and horseradish root and lay the symbolic place for the one who cannot, through adversity, be with us. These are rituals of remembering, the symbolic spilling of drops of wine from our glasses to represent the diminution of our happiness in recalling the affliction of the ten

163

plagues. Resolutely we recite the words, '*Dam, tsfardeya, kinim ...*' '*Blood, frogs, lice*', liberally accompanied by a cacophony of grunts and mutters from Chicken, who is nesting underneath the festive table.

The festival has arrived at just the right moment. For a while I've wanted to discuss various things with the family, to do with our relationships to town and country. This evening – the very occasion when, we've been taught, we must discuss our history and remember the nature of our place in the world – seems a fortuitous moment. (Although we don't need scriptural encouragement to talk.)

When thinking about cities and their ways, I've always accepted the prevailing idea, an extended mythology that has as its first principle that to be Jewish (or Jewish and of European origin) is to be utterly urban, that it was always so, or if not always, at least since the long-ago days of the Dispersion, and that if it wasn't so some thousands of years ago, it is now, an existing state of almost genetic certitude, immutable and for ever, an indissoluble part of the way things were and are, as definitive in us as the fabric of our selves. It doesn't take long to realise that this mythology can be swiftly blown apart by a glance at history. (Even as I write, a cousin is, with the kind of patience and assiduity I could never muster, tracing the footsteps of the family back into the nineteenth century and further, to the villages of Latvia and Lithuania where, I assume, our families did much the same – as far as occupation is concerned – as other rural and village families in the Baltic countries in harsh and difficult times.)

Many Jews lived in the cities of pre-war Europe, but many didn't. There are plenty of accounts and photographs of rural Jews, of grain merchants and woodcutters and farmers, as horny-handed and stereo-typically rural a bunch as any to be found anywhere on earth. Many lived, if not in rural settings, then in the small towns and villages scattered across the continent, living small-town lives sharply different from those of the capital cities we have come, wrongly, to associate most with European Jewish life: Berlin, Vienna, Prague, Budapest.

I've discussed it before with friends and family, and we have come to the conclusion that it's a matter of generation, and that the brief – in evolutionary terms anyway – history of Europe's troubled relationship with its Jewish population has defined us, often wrongly, as completely urban and that much of the perception, even our own, is based on observations, partial at best, of the history of the twentieth century.

We talk over the strange fare that characterises the Seder, the hard-boiled eggs in salt water, the charoset – the mixture of chopped fruit, nuts, cinnamon and wine that symbolises the mortar used in the making of bricks which was the Jews' task in pharaonic Egypt – and about my father, who died while I was still young, from whose loves and inclinations I can trace my own. The stories he told were always about cities, about New York or Chicago or London in the 1930s, stories of the blackshirts and the battle of Cable Street, of the Blitz and the blackout and shelters and air-raid sirens, the kind that they still tested from time to time throughout the years of my childhood. Fact and fable were often indistinguishable, to me at any rate, but somewhere, unspoken but present in the narrative, was the implication that his knowledge was gained through personal experience, which, given the momentous events he was usually describing (and the startling disparities of dates), it never, I think, was. When he told us once that he remembered standing in St Petersburg station during the Russian Revolution, I believed him, credulous child that I was. He had a romanticised attachment to the history of his time, and was fascinated by the United States in the era of prohibition, by the Chicago of Al Capone and the literature and music of a more exciting age. (While I remain in doubt as to my father's personal involvement, for a long time I looked at people carrying violin cases with a sharp, suspicious interest.)

In his loves and his deepest interests, my father was urbane; in his love of bookshops and clothes shops and silk ties, of hotels and restaurants and travelling to cities. He loved cars, too, and delighted in all sorts of gadgets: Minox cameras, odd and innovatory household gadgets: rumbling contrivances which were meant to peel potatoes but didn't, ironing machines with slow-moving, creaking rollers, juice extractors of vast, imposing solidity, long before such things became popular. He loved flying, and so we flew in the days before many people did on exciting propeller planes. My sister and I, often the only children travelling, were spoiled and given picture books and pencils and the badges and certificates they gave to people who had crossed the Equator although we hadn't – we'd only crossed the Channel. Airports then were sedate, unpopulated places, filled with the echoes of disembodied announcements and people who, helpfully, told you where to go or asked if you were lost. Europe was small then, a cluster of broken countries still healing or recovering, off-centre anchors for an uneasy continent.

The places we visited most were cities. I look back with amazement at the way we travelled every summer like some sort of minor royal progression to two or three cities on the way to a place with beach and sea, spent some time fortifying ourselves with sunlight and clear salt water against the challenges of a Scottish winter, then took the reverse trip home. We went especially to Biennales and Triennales in Milan and Venice, and to Rome and Amsterdam, Brussels and Paris, for my father to look for new designs for the furniture shop he ran. New York was his favourite place. I didn't go there first until much later but when I did, I felt that I was visiting a memory. He died when I was thirteen but even now when I go there, it's with a sense of expectation, as if I'll find him there, as if I have an appointment to meet him when we'll stroll together while I tell him about the progress of my life.

The thought of my father is not only one of loss, but of explanation. He was born only a few years after his parents had travelled, as so many did, from Lithuania to Scotland during the waves of emigration of the latter years of the nineteenth century, emigration of both economic and political necessity which took vast numbers of Jews from the countries of northern Europe, from Poland, Latvia, Lithuania and Russia to Britain, western Europe and the United States. His family arrived and, like virtually every other Jewish family both here and in the United States, settled in the city of first arrival, in their case Glasgow, to begin the immediate search for work and housing. Family life, community and religion itself imposed precise, exacting demands – there had to be places to buy kosher food, a synagogue to hold the services which may take place only when there's a *minyan* – a congregation of no fewer than ten men – and provision for all the other minutely dictated practices which makes Judaism at its more observant reaches the all-encompassing, total, every-minute-of-every-day, you're-not-going-to-forget-it-even-for-a-second kind of religion it is. Off the boat you got, at one or other of the ports of Britain or at Ellis Island in New York, and if there wasn't a relative fortuitously pre-established to collect you, you headed towards the nearest Jewish community or welfare organisation and wherever it was, you stayed, whether in the East End of London, the Gorbals or the Lower East Side of Manhattan. This was and is the experience of all immigrants, all migrants, all migrators, the journey of fear and hope, with, at the end of it, the earnest desire for any small and welcome element of the familiar.

One of the rare exceptions was the family of *New Yorker* writer Calvin Trillin, who, in an article in 1994 called 'Messages From My Father', writes of his father's arrival from the Ukraine at the age of two, in St Joseph, Missouri by way of Galveston, Texas and ponders the question of how his family, in their escape from the czar, ended up not in metropolitan New York as did most East European Jews, but in Texas.

Calvin Trillin's theory is that stubbornness dictated this unlikely settling. Having argued with a friend about whether you went to New York or Texas when you got to America, his grandfather headed for his suggestion, Texas, because he preferred to travel two thousand miles out of his way than be proved wrong. (Calvin Trillin's father's sole memory of the Old Country was a vague one of getting a foot stuck in mud.)

Arriving in a new country, and more particularly, arriving poor in a new country, places you uneasily in the society you join. You may know very little about it, not only about the laws but the rules, the social and political subtleties learned only with the confidence of time. If you're not actually afraid, you're watchful. Of some things you may indeed be afraid and so, often, you choose not to stray outside the small enclosure of your known environment.

You may have been rural, but you are no more. You may even be afraid of the countryside, afraid of what you don't know about who owns what, about how you might find out, how you might become entangled in some elaborate, Byzantine system of class and rights and possession. What might you do there, anyway? There's no time to go there, and no reason, no place in an organised countryside owned, managed, farmed, lived in by people who might not welcome you. Often, you know from your own life of the innate, unbreakable conservatism of the countryside. Immigration doesn't make you new, it alters your view of the past. It changes how and what you remember, makes you live differently, especially if what you've left is precisely what you don't want to reproduce in the possible new world of your future. You may wish to forget. Stay where you are. Stay close. Eat the food you know and shop in the shops you know and carry out the remnants of the life you left, the life you both do and don't want, a life of both past and present, and know that your children will look differently on it and that, if you don't talk about the past, it doesn't mean that it's not there, that it won't be disinterred, if not by your generation then by the ones who follow, the ones who carry the desire for connection, the fascination but no longer the pain. 'For those living abroad, the clocks stop at the hour of exile,' the Russian writer Alexander Herzen wrote.

This evening we read the words of the Haggadah: 'You shall not oppress a stranger, for you know the feelings of a stranger . . .'

We are all poised differently in our relationship to the countryside, feeling in ways explicable and otherwise the sense that as immigrants – even second- and third-generation ones – that the countryside is a place apart, a sense not easily dismissed, one that prevailed in my childhood and for all I know, still does. The move from country to city has been seen as both adventure, a promise of a better future but also the distancing from innocence, one that, in the case of the Jews, mirrors the move to the Diaspora, which in itself mirrors the flight from Eden.

In an essay, 'Dancing: A Grand Canyon Saga' published in 2002 in the anthology *At Home on This Earth*, Evelyn C. White writes of the African-American relationship with countryside and wilderness. During the rafting trip to the Grand Canyon she has undertaken in an attempt to overcome her own fears of the outdoors, she tells the river guides about the psychological role of water in the history of African-Americans, a legacy of forced transportation by ship, of segregated swimming pools and drinking fountains, the fire-hoses used against civil rights marchers, and of the lasting memories of the dangers of being hunted down by Klansmen and bounty hunters in open country. Another African-American White meets at a waterfall in the canyon best expresses for her the summation of the emotional pain and social expectation that prevent black people from being at ease with nature. Everything in their thinking, he suggests, leads them to feel that to be in the wilderness is a step backwards, a return to the primitive, and is against every lesson they have learned about progress and getting ahead. Describing the delight of her journey while urging others to rethink their attitudes, White celebrates a triumph over her fear of the outdoors and her exhilaration in achieving her aim of '"getting tight" with Mother Nature'.

Among the few non-urban Jews I've encountered were the inhabitants of the kibbutz where I lived for a time, a place set in what at least in theory was countryside, but felt more like a misplaced adjunct to a middle-class Berlin suburb at an indeterminate date before either of the twentieth-century world wars, with the addition of heat and scorpions. Established by German Jews before the war, it was well ordered to the point of obsession, oddly at variance with the general and particular chaos of Israeli life outside it. The circumstances of the lives of the older inhabitants seemed never quite to unite two distinct and distant worlds, one of farmers and agriculturalists with their discussion of fields and cowsheds and tractors, and one of bookshelves arrayed with a breathtakingly comprehensive selection of the classic literature of Europe, volumes of the Greek plays cosily abutting the works of Schiller, Zola and Heine. I always wondered if even thinking that indicated my failure to transcend traditional views of both my co-religionists and rural life.

There was a lot of outdoors there. I remember self-conscious 'nature walks' that weren't obligatory but near enough to feel they were; hebdomadal, Shabbat occasions when a large and bossy man called Menachem would march an invited group across a hillside, questioning them on the way about their knowledge of plant, tree, animal and bird. (To be fair, some of the knowledge stuck. I can remember one or two plant names more easily in Hebrew than in English. Likewise, for some reason, the word for silage. Always useful.)

I think of those walks, as I have done in the decades since, as uneasy and uncertain now as I was then about their purpose, about whether or not they were as much an exercise in attaching names in order to consolidate ownership as of learning about nature. Echoes and shadows remain from the time, the seemingly unending questions of possession and dispossession which have haunted my mind, unresolved, every day for all these years.

Allied to the perception of urban-ness, or possibly as a result of it, there was and may still be the contemporary belief (accepted by Jews as well) that Jews in particular are, in some profound though inexplicable way, detached from the world of the countryside, exempt from the necessities and desires others seem to feel to take part in activities which will oblige us to engage with anything outdoors. I question this, and worry that I might be judging harshly, or wrongly, an entire culture. I've already carried out a small straw poll among friends. The question posed is this – is it true? The respondees are unanimous. It's true. (The study isn't scientific. My representative sample isn't a representative sample. It is people I know and now, it's my family and between us all, we come up with one person who lives in the country and it occurs to me that if everything in the religion isn't specifically designed to prevent its adherents from straying inadvertently into the open, then it has happened by way of chance. I consider the general attitude towards camping. Camping? To make oneself uncomfortable, through choice!)

It's not that there weren't and aren't many distinguished Jewish zoologists, ecologists, ornithologists, biologists, scientists of every hue and variety involved in the natural world in all its manifestations, it's just that it seems outdoors is more easily acceptable when it's decently cloaked in the respectable, admirable cloak of academe.

I was brought up in an air of protected timidity familiar to me from the lives of friends and family too, and I wonder now if the events of the war, for my generation only at our backs, encouraged this care and fear. Some years ago, I listened to a sketch on the radio about a British Asian family at a picnic in the country and laughed (and winced) with familiarity at the panic and consternation caused by the appearance of a bee. Culturally determined these things may be, or the universal experience of immigrants, but they shaped all my future attitudes to the world outside.

In his 2011 PhD thesis, David Dee examines the role of sport in the lives of Jews in Britain during the nineteenth and twentieth centuries, questioning wider social attitudes and the part played by sport in the integration of immigrant communities. In documenting the contribution made to sport by notable Jewish sportsmen and women, Dee redresses the view of Jews as being non-athletic, reluctant sportsmen.

In spite of this noble roster of Jewish athletes and boxers and runners, for many Jews there still seems to be a sense of separation between body and, if not soul, then brain and that rightly or wrongly, when it comes to Juvenal's sensible dictum about the relationship between healthy mind and healthy body, we've always held the view that the former has to be specially nurtured while the latter can, with suitable quantities of appropriate feeding, be left to get on by itself. If my generation (and, for all I know, subsequent generations) were less than enthusiastic about competitive sport at school, it might have had something to do with the fact that matches were played on a Saturday, which would have conflicted with our religious duties. But then again, it might not have.

Recently, I read a piece written by Jon Ronson on his love of running, in which he describes an incident in New York when, finding himself being chased by muggers in a dodgy district, he managed to outrun them. He was, he says, only the second Jew to manage such a feat, the first being Dustin Hoffman in *Marathon Man*.

It's growing dark as I open the door to welcome the silent, invisible presence of Elijah the prophet who, by tradition, visits every Seder. After he has been and we have watched the wine in the glass filled for him tremble slightly in the draught, I escort him out. The moon is full as it always is on the first night of Passover, almost golden in the streaked lapis sky, and as I close the door behind the departing sage, I know that the swifts will be back soon.

Angels in the Streets

On an evening uncharacteristically warm for late April, I join a 'bird walk' by the River Don where it flows through the city at Seaton Park. The walk is part of an artists' residency at the university in preparation for a conference to be held in the summer on conservation and the conflicts that arise from it. Andrew, the person leading it, is an anthropologist, a specialist in birdsong and its role in human life and perception. The evening is warm and golden-green and heavy with spring, an evening to make you remember what warmth is like, to give you, briefly as it turns out, the tentative, tenuous promise of summer.

On the way to meet the others, I walk through Old Aberdeen, through the cobbled lanes and streets, past the high, old stone walls and the buildings, most of which belong to the university, as they have since the fifteenth century. Soon it will be exam time, for me still a moment of personal phenology, the sight of cherry trees in flower connecting me into a never-ending loop of anxiety and hope. Above, jackdaws exchange *quorks* from the medieval roofs and chimneys in the High Street. The gardens of the gracious eighteenth-century mansions and the low stone cottages at Wright's and Cooper's Place are green and bright and flowering. Sun glints from the rows of windows of the modern blocks which seem less an intrusion than just another layer over the bedrock of time. In one of the houses I pass in the High Street, William MacGillivray, the great Scottish ornithologist, was born. One of the many distinguished natural historians of the city, he was famous for – among other accomplishments – the extraordinary expedition he made in 1819 when he was 23, having decided that, in order to visit the Natural History Museum in London, he'd have to walk there:

> In London city there is, I am told, a great collection of Beasts and
> Fishes, of Birds and other flying things, of Reptiles and Insects –
> in short of all the creatures which have been found upon the face
> of the earth. I hither therefore shall direct my steps – because I am
> desirous of furthering my cognition of these things.

In the eight weeks it took him to complete the journey, he kept a detailed journal of the 800-mile hike. (His route was longer than it might have been because he made extended detours on the way.) Some of the observations he made during the journey, entirely aside from those finely detailed descriptions of plants, may well explain his later reputation for being contentious:

'I had been led to believe that the English peasantry had disproportionately small legs but this is not the case in the counties through which I have passed.' 'The English women are smarter and more cleanly than the Scotch,' he mused, these benefits appearing, fortunately, to have affected his future decisions less than the overwhelming impressions gained during his visit to the Natural History Museum, experiences that impelled him ever onward to 'furthering his cognition' and to continue his studies in natural history. In 1841, he became Professor at Marischal College at Aberdeen University and subsequently wrote, among many other books and treatises, *A History of British Birds* and, as a result of his collaboration with John James Audubon, the *Ornithological Biographies*, a sequel to Audubon's *Birds of America* because Audubon, a French speaker, felt that neither his English nor his knowledge of taxonomy were equal to the task of writing anything so demanding on his own.

This evening, St Machar's Drive is busy and loud with traffic but The Chanonry, a cobbled street of Georgian houses behind high trees and walls, is quieter, with its ancient air of academic and ecclesiastical calm. Students wander and chat in the sunshine, making their way to their halls of residence on the other side of the park below.

We meet by the cathedral – Andrew, my fellows on the residency and some others – and as we stand waiting, a woman I don't know begins to talk about pigeons.

'I hate them,' she says, decisively. 'There are lots of them round us. Nasty, dirty things. They've got dirty feet and they shit all over the place.'

'There are worse things,' I say mildly, 'than bird shit.' (I speak as one who commands rare expertise in the subject.)

'Is that your motto?' Andrew says, and in his question I'm faced with an inevitable truth. I hadn't thought about it, but it is. There are worse things than bird shit. I will have it engraved upon my coat of arms. There will be a number of chosen birds, beasts and smaller life forms with these sage and weighty words inscribed at the foot because, inelegant though they may be, they hint gently at the misperceptions of the non-bird-owning world. (What can be worse than bird shit? Where do I begin?) They will serve as a necessary reminder of misplaced scales of value and certain inarguable concepts and will hint too at that laid-back, laissez-faire quality by which I would like to live my life.

As we set off in our assembled group down one of the steep paths leading to the park, I think of the question posed by Barbara Allen in a splendidly comprehensive book on the subject, 'Who can doubt the beauty of the pigeon?' and I remember the little pigeon I found in the snow.

The trees this evening glitter with new leaves, so new that on the canopies of the taller trees, they are still just opening buds. The foliage of daffodils wilts among the roots and grass. Walls of shrubs in the formal gardens, potentilla and viburnum, have been recently pruned, their margins blunt and twiggy. In the planted borders the brilliant flowers of summer aren't yet in bloom although in a few weeks they will be. Fresh red leaves have begun to grow from the low stems of the rose plants in the neat and elegant flowerbeds. Between them, the tidy grass paths are dotted with crows who dig busily, picking and delving for leatherjackets, leaving small scatterings of moss behind them (as the blackbirds do at dawn in the grass in my garden) like tiny avian gardeners. Across the stretch of grassland, beyond the tall sycamores and wych elms, the beech and chestnut, children are swarming over the red engine in the playpark as my children used to do on warm days of summer.

Writing of the bird life of this park, Peter Marren in his 1982 book *A Natural History of Aberdeen* describes some of the vicissitudes in the lives of park birds, of the willow warblers mown down after nesting among daffodils; of redpolls, goldfinches and a breeding pair of lesser whitethroats – found in 1977, the first ever to be seen north of the Forth – all losing their chosen habitats in overgrown bushes to a park-keeper's zeal.

We walk towards the banks of the Don, past a sign prohibiting barbecues. On the grass under the trees, people are setting up barbecues and laying out picnic rugs. This may be one of the only warm days of the year. The river is perfect this evening, glittering sunlit on this, its final stretch, full from the winter's snow, only a mile or so from where it will open out into the sea beyond our sight, eighty miles from its source in the high ground of the Cairngorms, east of Corgarff.

The path we take rises above the water, through the steep banks of beech and chestnut. As we walk, Andrew listens and interprets, as if he's identifying and translating from endangered, arcane languages, listening to voices from other cultures, which is what birdsong is. He concentrates on a chorus, picks out the individual voices and names them: dunnock, warbler, blackbird. He reminds me of a conductor explaining a detail to his orchestra, someone you could listen to for ever talking on his subject. Below us, the water flows round outcrops and buttresses of rock, speeding and frothing then stilling into deep pools of illuminated green. We walk, listening, watching in the warm spring light. White-flowered stitchwort grows in the grass, wild garlic and cow parsley, wood sorrel and bluebells. It feels far from the city, from the halls of residence visible from along the path, from the main roads just out of sight, from the hectic life surrounding us.

This evening, I'm expecting a visitor and have to leave the bird walk early. I run back along the path by the river, through the park, through air quickening with insects. (Later, Andrew and the others will be surrounded by a cloud of them, ones they can't identify.) Below me, a heron lifts lightly from its rock and melts into the shadows on the river. By now, the barbecues are well under way on the flat grass of the banks, the packs of beer unloaded, scents of cooking food spreading in the air. As I pass, a young man with a camera and an unidentifiable accent stops me and asks if I'll take a photo of him and his friends. I do, and on the way home I think of the life of photos of groups such as this one I look at now, in books about poets or artists, the life of the past in elegant gatherings or assemblages of the bohemian and raffish, all locked behind the vitreous stillness of time.

Over the next few days, the discussion on pigeons stays on my mind. I've been wondering why someone sufficiently interested in birds to spend an evening listening to their song might be so hostile towards pigeons. What is it about those birds that arouses contempt and dislike? I'm still thinking about it when, a few days later as I'm walking down Union Street, a pigeon flies up suddenly before the approach of someone's feet from the pavement where he has been picking, colliding with my arm in a brief whirr of feathers and flapping. A young woman walking near me screams and clutches her companion, shrieking, 'Oh my God!' A handsome, dark-feathered bird, only slightly town-worn, he seems undamaged by this minor collision (as I am) since he stops only for a moment on the edge of a rubbish bin before flying down to resume his task. The two girls continue to clutch one another anxiously in the background, exchanging sounds of mutual alarm and consolation, and as I walk away, I realise that this may be the most serious encounter they have ever witnessed between a human and another species, and pigeons one of the only bird species they ever see. What do they know of the origins of pigeons, of their lives, their emotions, these birds loved and kept by artists and kings and people fascinated by their qualities of flight, the bird of Noah, accompanying us closely through historical and mythological time, at the heart of our iconography, religion and beliefs, possessors of navigational abilities we don't yet fully understand and of capacities we haven't even begun to know? Do they think of pigeons as anything but 'rats with wings'? (A term used first, it seems, by Woody Allen in his film *Stardust Memories*. Misguided and inaccurate though it is, it has supplied a rationale for pigeon detractors everywhere.)

I try to imagine the streets without these birds, the pigeon family behind the plastic sign of a newsagent in Union Street who I've watched for many years, taking an almost familial interest in the seasonal sounds of hatching and the first appearance in public of the young, the ones I see nesting in the skylights of the vennels near Castlegate, the ones who stand proprietorially in the glassless windows of an abandoned warehouse, their lives and presence another of the markings of the day and year.

Often it seems as if there's a lamentable apartheid when it comes to the one member of the 300 or so species of the avian family

Columbidae with whom we are most familiar. At least part of the difficulty lies with the words 'pigeon' and 'dove'. Seen wrongly as separate species with different habits, inclinations and appearance, city pigeons, most often dark grey, blue, slate, brown or black, emerge from the confusion at a great disadvantage, bearing with them our every prejudice towards urban life. Doves, generally thought of as smaller than pigeons and white, are regarded as domesticated, flying from neat white dovecotes, surrounded by a nimbus of purity, innocence and spirituality. But town or country, grey or white, homing or racing, they are the same bird, *Columba livia*, with the same line of descent from the wild rock pigeon.

Pigeons, with their 34-million-year-old roots in the Eocene and Miocene, were here long before we appeared, with time spreading widely over the surface of the earth where some species still thrive and others are endangered as a result of the environmental factors that may affect any species. As varied in appearance, name and habit as any family of birds, they can be as different as the Victoria Crowned pigeon of New Guinea with its frothed crown of blue, or the Luzon Bleeding-heart of the Philippines, splashed across its breast with brilliant feathers of blood-like red.

As with every other 'urban exploiter', the worldwide urban populations of *Columba livia* live on what we leave behind, and when we deem their presence unwelcome, we seek to have them removed or destroyed. Too easily we forget their many roles in our lives over the centuries as providers of both food and fertiliser or as vital messengers sent aloft from sieges, over trenches and battlefields, soaring with avian

detachment over our *Homo sapiens* exploits. They have provided for us
the embodiment in avian form of the manifestation of the Holy Spirit
(although I have always been sceptical of the appropriateness of this
particular role, having witnessed some very un-Holy Spirit-like behaviour from my doves), a source of pleasure in their appearance, in their
ability to fly at speed and to return, and in the calm fascination of their
company. Can anyone visit a city and, on seeing a neat flight of pigeons
turning and soaring in the late-afternoon air over the Upper West Side
of New York or over the lovely mansard roofs of Paris, fail to imagine
the pampered circumstances from which they probably came, the
devoted attentions of the person who has bred and nurtured these
birds, who has worried about their diet, their health, their well-being?
For above you as you watch may be the elect, possibly the most fortunate creatures on earth, the inhabitants of a pigeon-fancier's loft.

Among the many who have loved and kept pigeons (Matisse, Picasso
and Mike Tyson being notable pigeon-fanciers), Charles Darwin
devoted time and energy to the ninety pigeons of all the breeds then
available that he raised in his dovecote at Down House. Convinced
that, despite morphological differences, these birds were all descendants
of *Columba livia*, he kept and drew pigeon skulls, noting that the variations between them were such that they could be thought to belong
to separate species although he knew they didn't. If selective breeding
could bring about such changes in a short time, the possibilities for
change over millions of years, he realised, were clear. Inviting his friend
the geologist Charles Lyell to visit him, he wrote: 'I will show you my
pigeons! Which is the greatest treat, in my opinion, which can be

offered to human beings.' (An inducement I've been known to use myself.)

Rather like 'fancy' rats and their shared origins with the wild variety, even the many 'fancy' pigeons, domesticated, specially bred birds of fabulous appearance and astonishing habit are *Columba livia* – yes, even the Rzhev Startail Tumbler, the Franconian Trumpeter, the Flying Tippler and the Spangeled Magpie Purzler of Satu-Mare.

Redoubtable 'urban exploiters' pigeons may be but they are also renowned and remarkable travellers, able to find their way unerringly, navigating their journeys and orientating their direction by using visual information from the stars, from the sun and polarised light, steering by mountains, rivers or roads, by scent and sound and by their still unexplained, hitherto inexplicable use of the earth's magnetic field. In spite of intensive study over decades, the mechanisms of magneto-reception – the means by which birds appear to be able to detect and employ the earth's magnetic field – remain unknown. For a long time, it was thought that cells of iron magnetite crystals in the upper beaks of pigeons were involved but recent research at the Institute of Molecular Pathology in Vienna has revealed them to be macrophages, cells active in the immune system but not in direction-finding. In a separate study at Baylor College in Texas, it was discovered that 53 cells in the brainstem of the pigeon react both to magnetic-field strength and direction, and although work is being carried out into the role of a light-sensitive protein, cryptochrome, in birds' eyes, or the possible location of magnetoreceptors in the ophthalmic branch of the birds' trigeminal nerve, it is still unknown exactly how they find their way.

If we don't yet understand the physical means controlling homing behaviour, one of the social mechanisms is pair-bonding, which will lead a bird back to its mate. In common with 90 per cent of birds, pigeons are monogamous, which in avian terms means that one male bird pairs with one female during one breeding season. Many birds are either socially or genetically monogamous – the former suggests that in forming pair bonds for breeding and rearing young the bond is not exclusive, while in the latter, it is. Part of the explanation for a behaviour relatively unusual in the natural world is that since male birds are able to carry out the same incubation, feeding and care of young as female birds, monogamy with its investment of parental care increases the chances of breeding success. While it was once commonly believed that many avian species were monogamous, the advent of DNA studies has demonstrated that belief and observation are not always correct, and that far fewer birds really are than those believed to be so – only 14 per cent of those tested. In a chapter tellingly entitled 'Idyll and Mayhem – the Sex Lives of Birds', biologist Colin Tudge in his book *The Secret Life of Birds* reveals both the inaccuracy of some of our cherished ideas and the often astonishing sexual behaviour of a variety of birds for whom 'extra-pair copulation' is very much the norm. (Anyone who reads it will never look at robins, dunnocks or cliff swallows the same way again. Although Columbidae mate for life, they too have been observed to engage in extra-pair copulation.)

We co-opt doves to be our symbols of love and fidelity, so should we be surprised by their showing love and devotion? When Nicolas Malebranche expressed his opinion of the joyless, feeling-less, fearless insentience of animals, clearly he couldn't have countenanced ideas of attachment, fidelity or cooperation existing in any but the human world. Anyone who observes birds has witnessed manifestations of emotion. In his book *What the Stones Remember*, the Canadian poet Patrick Lane describes the reaction of a hermit thrush to the killing of its mate by a car:

> The male bowed to her and cried a new, impossible song. It was not the song of warning and not the song he sang when he was trying to win her attention this past month ... This was different. It was a trilling that stuttered into wild squawks. He grasped her wing in his beak, and pulled it a bit, urging her to fly. When she did not respond, he flew up once more, threw his head back, and sang again. It was a song of grief. There is no other way to describe it. I have heard the same song in the streets of Peruvian villages, in the mountain towns of Mexico and Colombia, Xian and Rome. I have heard it in my own land. It was a cry to God. It asked for impossible answers. It was the song that precedes Kaddish, precedes mourning.

In *Pilgrim at Tinker Creek* Annie Dillard, writing of the experience of grief in animals and what appears to be a capacity for emotion similar to our own, in a question of transcendent unanswerability, asks: 'What creator could be so cruel, not to kill otters but to let them care?'

Anyone who has seen the reaction of a dove to the loss of its mate to a hawk couldn't doubt their ability to feel. I've read of the mourning

rituals of magpies. I have seen the fear, the *timor mortis*, in many creatures, a fear that may transcend every boundary, recognisable in species
other than ourselves for it is a fear common to us all.

Might knowing a bird's capacity for feeling prevent anyone from
calling it 'a rat with wings'?

When we were first here, very few wild doves came to the garden but
over time, collared doves began to appear as they moved steadily north
and west from Europe and the Middle East, expanding their numbers
in Britain from four birds in 1955 to their present populous ubiquity.
Streptopelia decaocto, lovely birds of soft fawn grey with narrow torcs of
black and white round their graceful necks, the high arc of their wings
and the beauty of their faces make them look as if they might have
flown from the pages of *A Peaceable Kingdom*, a Shaker abecedarius I
have, the book of animal rhymes with which Shakers used to teach
their children to read and one I used to read to my children, exquisitely
illustrated by drawings of gentle lions lying down with lambs and
woodpeckers busily eating the handles of brooms. Every day, I watch
them fly in to land with a straight-bellied drop, a last-minute lifting of
their angled wings.

Over the years, *Columba palumbus,* wood pigeons, too began to be
seen in the garden, padding over the lawn with their admirable solidity, their round, surprised, solemn eyes. No less beautiful than collared
doves, they are feathered in the palest of pink and lilac, skimmed over
with a fine glaze of amethyst, their heads of slate grey, their wings lavender and turquoise. Their numbers have increased in towns, possibly
because of changes in farming practice, in types of crops and timings

of planting. (According to the RSPB, the Royal Society for the Protection of Birds, collared dove numbers have increased by 370 per cent, and wood pigeons by 850 per cent since 1979, making them both among the most common of garden birds.)

I put out food for them all but there is still the occasional food-inspired altercation between resident and non-resident birds. (My doves, already well fed, have always enthusiastically embraced the concept, best known from Thomas Mann's *The Magic Mountain*, of 'second breakfast'.)

These days, the doves I've kept for over two decades stay inside their doo'cot in the garden more than they used to, mostly from their own choice. I open their door but they won't go out. They have plenty of room in which to live but every season is sparrowhawk season, and my doves are fearful. I don't know if age increases their fear and makes them timid. I feel more protective than I did when they were young and easily replaced. For many years, they reproduced so relentlessly that the population problem was one of over- rather than under-abundance. We gave young doves away before we learned the trick of egg removal. Their fecundity made it easy. I was happy to be an unsentimental keeper of semi-wild birds, believing that I favoured no species in my admirable recognition of the rights of all, or at least of Columbidae or Accipitidrae, but age, theirs and mine, has, if not altered my general view, then affected my behaviour. If I could fly with them, usher them protectively into the sky while keeping a fussy maternal watch over them, shout and rush and frighten passing sparrowhawks, if I could call them in, urge them in a glittering white and blue flock back into the fastness of their house, I would. I would welcome them in their ancient pairs and shut the door behind them but on every fine day when I open their door, I know that in the moment they step from their house, launch themselves ecstatically from their platform into the endless air, they are alone. On a day last summer after heavy rain, I let them out to fly in the sunshine while I swilled out and dried the results of the night's storm from their floor. Within half an hour, I saw from a top window a familiar splash of white on a neighbour's lawn, the red heart at its centre, the spill of feathers and blood.

Increasingly, on the days I try to usher them out into sunlight, one party or another will step reluctantly out onto the platform and look around at the prospect of the skies, before turning back into the safety of the dove-house. I no longer shoo them out routinely every time I clean their house as I used to do. I realise that I have to respect the foibles of age.

My granddaughter Leah and I go to Venice. In spite of her long association with Columbidae, every time we go there she enjoys feeding the pigeons. (There is, extant, a photograph of my sister and myself in childhood, younger than Leah is now, feeding pigeons in exactly the same place in St Mark's Square, strangely empty, in the far-off days before Venice became a major tourist venue. There we are, grinning in blue and white checked cotton dresses, surrounded by the progenitors of the pigeons Leah feeds.) Whenever we've been, I've thought briefly about taking supplies of bird food but have never wanted to be found transporting pigeon grain across Europe and possibly unmasked as a serial bird-food cheapskate. Besides, I didn't want to deny the pigeon-food sellers their living. Obediently, we bought the bags of food for sale and Leah handed it out and we were consequently surrounded by birds. They stood on our heads, our shoulders, our arms, our hands and displayed, I have to admit, more trust and confidence than my lot have shown me in the twenty-odd years I've fed, cleaned, valeted, nurtured, acted as midwife and matchmaker, worried about and yes – even loved them. (There is a lesson there somewhere.) While I was happy to stand around among pigeons, it did feel rather like a busman's holiday. Crowds of Japanese tourists giggled and squealed and photographed each other covered in pigeons. When it was time to go we picked the pigeons off one by one, held them tightly in our hands and like some putative bird-loving saints making offerings to a beneficent heaven, flung them aloft to soar majestically above St Mark's towards the Museo Correr until they turned and descended to alight on the next group of tourists. St Mark's Square always looked remarkably clean, considering. Someone, I realised, must have put in a great deal of work.

In John Berendt's *The City of Falling Angels*, he describes the experience of coming unexpectedly upon the early-morning trapping of Venice's pigeons, as four men catch batches in a net, an operation carried out clandestinely, he discovers, because all attempts to reduce pigeon numbers have been vigorously opposed by animal rights organisations. On being questioned, the trappers tell Berendt that the pigeons will be taken away and released but on seeking out the chief instigator of pigeon-removal, one Dr Scattolin, Director of Animal Affairs, he is told that in fact the pigeons will be chloroformed. (Of course they couldn't be released. They'd only return speedily whence they came –

they're homing pigeons.) Dr Scattolin explains the predicament: the overcrowding, the numbers (at the time of Berendt's writing, some 120,000), the possibility of disease spreading, the less sanitary habits of pigeons. More than that, he explains the role of tourists in maximising the problems of numbers. Tourists like being photographed with pigeons, holding pigeons, having pigeons standing on their head. They buy bags of grain to feed them and a well-fed pigeon will breed all year, seven or eight times. They have tried peregrine falcons, who killed only one pigeon a day each and produced excrement even worse than that of the pigeons. Neutering would be too expensive and contraceptives in their food don't work. When Berendt suggests banning the pigeon-food vendors, Dr Scattolin tells him sadly, resignedly, that the selling of corn is so lucrative that the vendors can afford to pay huge sums to the municipality for their licences and so, since the tourists like the pigeons and the municipality likes the money, there is no solution.

But then, in 2008, the law changed and the pigeon-food vendors were prohibited from selling food and their licences taken away. The pigeons had become too much of a nuisance and were destroying the fabric of already crumbling Venice. But of course, the pigeons are urban exploiters. They don't go away. For the moment there are fewer but they're still there.

(Some European cities, in dealing with their pigeon problems, have established municipal dovecotes where pigeon-feeding and numbers can be controlled. Egg removal and regular cleaning have reduced pigeon populations and kept cities clean more effectively than lethal controls, which tend to increase populations.)

I have on my desk one of the small grey paper bags in which the corn used to be sold. Produced by *I Venditori di Grano di Piazza San Marco*, it is now the relic of another age, the pigeon age. On it is printed the motto: 'The joy of beauty, art and history must be defended.' On the other side, mysteriously, in seven languages, the exhortation, 'Always dress decently!'

SPRING TO SUMMER

SPRING TO SUMMER

April 26th

It's no longer warm. April progresses towards May in chilly, uncertain brightness. The bird is still here. One morning, I see him leaping glass-ward, his beak filled with fine strands plucked from the dry stalk of clematis, the one I thought hadn't survived the winter. The single green shoot has been joined by the many others that sprout improbably from the shredding, shedding stalks.

Inside the house, Chicken finishes nesting. She moves in stages back to her own house and as she does, I remove a few pieces of newspaper from under the table, hoping she won't notice. She doesn't. Within days, the matter is over for what I hope will be another year. I wonder if the great tit will stop his window-tapping soon.

May 1st

The first bird we ever got has died. He was 22 but looked remarkably little changed from the day we bought him when he was 11 weeks old. Birds have that advantage over us – it's difficult, if not impossible, to tell their age. Some birds attain their definitive plumage within their first year and that, apart from temporary changes brought about by moulting, is the way they stay. For other birds, arriving at definitive plumage is a long process – gulls take four years, eagles five as they pass through natal down and juvenile, winter and first-year and sub-adult plumages and all the other stages and names on the way to attaining their adult feathers. Birds live for a long time, and their longevity increases the potential sorrow of this unique personal relationship. Parrots can share an entire lifetime; rooks a span that covers decades.

This bird is Bardie, Bec's cockatiel, bought for her twelfth birthday. He has been my companion, a reluctant one on his side at least, since Bec left home. He has grown frailer over the years, more reluctant to emerge from his house, his glory days of power and influence dimmed to the occasional bout of persistent shouting. (He stoutly maintains his dislike of me to the end.) Until recently, he still greeted Bec with fervour and joy but when she left after a visit, he'd fall silent awhile, a silence often broken surprisingly by vehement if intermittent song. He refused to come out of his house if only I was there. Even with Bec, he grew reluctant. For a long time he has been like an old statesman, quiet with memory. I became more careful when I bathed him, making sure he was stable on his perch before spraying him with water. He enjoyed being bathed but, aware of his own frailty, became alarmed by the prospect. I would wait until sunlight struck the worktop where I placed his house, until the kitchen was warm enough to dry him swiftly afterwards. I kept the process brief. He still closed his eyes with pleasure. He had difficulty in manoeuvring his way about his house, holding onto his bar and perch with the proud, uncertain determination of the elderly.

At the moment I hear an unfamiliar sound and see him stumble, I recognise that this is different from his usual cautious progress. I'm on the phone to Bec when it happens and, as carefully as possible, warn her that all is not well. He is, as he has always been, her bird. As I approach him, I see the dulled look, that opacity of eye, the sign in a bird that something is no longer present. I suspect that he's had a stroke, some catastrophic change in his brain, and the next morning, after one last baneful, reproachful look, quietly he dies. Twenty-two years have passed in his small but mighty presence and now, I'm aware of the tasks I'll never do for him again, the feeding and bathing, the purchase of both necessities and gifts. I evaluate his life. He was my daughters' childhoods. He was my lesson, a beginning, the first small, brilliant aperture into the vast place beyond humanity. When I bury him in the garden, it's with the sadness for the passing of an ancient. Bec, who always suspected him of being the embodiment if not the reincarnation of Napoleon, passed the important years of her life with him, her later childhood, her teenage years. Even in the years since she left home, she has regarded him as the closest of companions. The death of this bird is a landmark for me, too and I'm aware that in the months that will follow, after his deficient house – much Sellotaped and glued and held together in whatever way possible to prolong its life to prevent him from having to endure unwelcome change – has been carried to the bin, after the corner where he resided magisterially for so long has been redesigned, items moved and pictures hung, all to fill a space that cannot be filled, I will become no more used to the redundancy of my own actions and words, to the automatic greetings, the one-sided conversation. I will never cut an apple without leaving a piece for him, forget to check that there is sufficient birdseed in the cupboard. These are the habits of 22 years, and I accept that there is a corner to which I will always turn, surprised, a small grey figure in the corner of my eye, that there will be no entering the room without the stifling of a greeting. Like Elijah, or the person who wasn't at the Passover table, he is here but he is not here. He has done as they all do and did, every bird, every creature with whom I have lived, left a permanent imprint on my life and brain. I accept that for however long I live in this house, every evening I will say goodnight to an empty corner.

May 3rd

And then this morning, the great tit too is gone. It's the day after I've found two great-tit fledglings dead on the grass. I can see no signs of violence. I'm sorry as I pick them up. For this moment, I see them and not the wider picture of their lives, the numbers, the losses and the breeding patterns, the forces that destroy not just them but their habitats, their food sources, the network of reliance upon which all creatures survive. I would rather they had been eaten than had just fallen or been left lying on the grass.

By now, I'm used to the bird. I miss his clicking. I hope he's merely occupied with nesting, a time that's precarious not only for small birds but vicariously for everyone who watches, nurtures, cares. Every breeding season, I fill the bird-feeders, put out nest material, encourage silently and watch and listen. I note the stages of nest-building, the preparation, the sitting and hatching and feeding. I know that the flight path to the blackbird's nest is straight towards the kitchen window then a quick left turn. When I hear the sounds of nestlings, I take up my binoculars and focus them on a tunnel in the leaves of the ivy, five feet away through the rat-room window. A yellow gape and a watching, infant eye fill the lens.

May 5th

I read a letter to a newspaper in which a man asks why the garden birds he feeds so assiduously and with such love refuse to approach him and are as untrusting as they are. What is he really asking? He knows that small birds expect little from humans and I suspect that his question is more profound. It's about the distances between us, the staunchless need and desire for those who wish to share the world equally with other species to be accepted into the lives and emotional ambit of other creatures. Who is going to answer him? Who is going to tell him what he already knows, that gratitude isn't part of the bargain, that small birds live brief lives of ever-present danger, that many of the passerines of our gardens are solitary, that the very behaviour of which he despairs is what keeps them safe, even to the minor extent that they are? (If it's relationship he's after, he needs a different kind of bird entirely.)

I recognise his feelings and our human need for acknowledgement, love or gratitude but how did we, as a species, come to want or need a response? In an old book I find lengthy instructions about how to persuade a variety of birds to land on one's hand. The endeavour takes a long time and endless patience. Wonderful though the experience may well prove to be for the human, might it be as rewarding for the bird? It succeeds only because inducements are offered and therefore the bond is one forged not through closeness or mutuality but from a dangerous kind of encouragement – the trading of trust for food.

We are so selective in our loves and our hatreds. We excoriate creatures for behaviour we can barely understand, invariably behaviour that ensures their survival. We admire them for qualities they do not and should not have. We express regret or even anger when they behave in ways that are true to their nature but not to ours. We condemn magpies because we witness what we deem wrongly to be their crimes in taking fledglings, but not robins, whose behaviour is highly aggressive towards members of their own species. Influenced as we are by image and by a long history of anthropomorphic portrayal of the natural world, we assign roles to creatures, demeaning them by casting them as small replicas of ourselves. We portray animals in caricature and the effect, as with the stereotypical portrayal of humans, is to deny them respect, to distance ourselves from their needs and nature. Upon larger species, ones not living among us, we bestow concepts beyond the idea of merely attractive, a subjective enough judgement in itself. We're partial, often simply because of the appearance of a species, or what we believe their characteristics to be. We divorce them from the context of the natural world and are apparently prepared to give more money to causes which promote the interests and future of creatures we deem to look as we wish them to look, or have qualities we believe them to have, than those who, in the course of the general destruction of habitats, may require our money or our attention more urgently. The 'unattractive' may be overlooked, however valuable their ecological role may be while big cats, pandas, the higher primates or the small and 'cute', the meerkats and lemurs may be designated 'iconic' or 'charismatic' (which makes them sound like Bill Clinton, or Rasputin) and we're prepared to donate or to campaign to save them from the fate that may befall any of us on earth. Often, research endeavours to demonstrate that other creatures are more like us than we think while at the same time the opposite is suggested, in attempts to attribute one or other piece of egregious human behaviour to 'animal' instincts which we excuse or smile upon under the guise of terms such as 'alpha male' or 'hardwired'.

I look at Chicken, who comes to stand on my knee after her bath. Under the light, the droplets of water shine along her beak, on the fine hair-like feathers on her face and head. She's like the bird on a Christmas card, frosted and sparkling, except you don't often find socially minded, intelligent, empathetic corvids depicted on Christmas cards. Only territorial, aggressive robins.

For all our misperceptions, is the relationship humans have with other species a universal human feeling, a need, innate and instinctive, to have a bond with the natural world, a need that has evolved over the history of the development of humans and other species together on earth?

In 1984, the distinguished biologist E. O. Wilson published his book *Biophilia* in which he suggests that humans possess an evolutionary predisposition towards a love of nature, as old as the fabric of ourselves, one he describes as 'seemingly an inborn trait', which, among other effects, gives us preference for certain types of landscapes as a consequence of our origins on the African savannah. The idea of 'biophilia' proved contentious from the beginning, arousing criticism from those who regarded it as sociobiology (a term popularised by Wilson himself) and therefore dangerous in its implications that human social behaviour is determined by evolution rather than by environment and social circumstance. Citing its unprovability, many scientists have continued to be critical, contending that if biophilia does exist as an inborn trait, it is one that can be suppressed, ignored or overridden by human need, greed or ignorance. In spite of this, the term has extended beyond its original concept and has been widely adopted and used by those who see it less as an evolutionary theory and more as a rallying call, an important basis for encouraging a greater awareness and care for the natural world and its inhabitants.

In the book, Wilson writes of being in Old Jerusalem and wondering about the native Palestinian plants and animals that might still be found under the olive trees of Gethsemane. Watching ants, he muses on the mention of them in Proverbs: 'Go to the ant, you sluggard! Consider her ways', and he wonders if the ants are of the same species, in the same place as those observed by the Old Testament writer (generally supposed to be King Solomon, that considerable commentator on natural history) and suggests that 'the million-year-old history of Jerusalem is at least as compelling as its past three thousand years'. It is. They're both compelling; worthy of our admiration, fascination, study and devotion; the lives and histories of both man and ant.

May 10th

It's been spring for a few weeks, if April and May constitute spring, if some fleeting, unexpected sunshine does, or even the sound of newly returned swifts screaming over the garden. The temperature has crept up as high as 16 degrees Centigrade. Spring has been as evanescent as it often is on this island of wind and rain clouds and indeterminate seasons.

I think of the death of the great-tit fledglings. I'm here all day at my desk, observing, turning my head in an instant to any sound or disturbance, running outside in my usual futile way at any sign of disruption but I know that it's in the very early morning that most of the dramatic events happen in the garden, before I'm around to see them. I worry that there's more I could do. Wake up earlier? Mount patrols? In the end, I know that there's only one thing I can do: accept. More than that; acknowledge. I cannot save anything in this world. I cannot defend anything from anything. (Child-rearing is the most ambitious enterprise in this direction that I have endeavoured.) I find baby birds drowned in the shallow birdbath in the garden. I lower the level of the water and feel excoriating guilt although I know that it is both my fault and not my fault. During the time when the small blackbirds are about to fledge, I discover infant feathers thinly arrayed on the

203

moss in the grass beside a curled white pile of blackbird innards. With this finding, I come upon the explanation for an outbreak of alarm-calling I heard in the early morning. Something will, having fed, live for another day. (Some weeks later, in the garden, under the guelder rose beside the pond that I now allow to grow wild, more bog than pond, I find a perfect pair of small black wings, still feathered, joined by a fine bridge of bone. The head and body are gone and what's left is already silvered by slime, decorated greenly by wisps of moss. I pick them up and bring them in and lay them on newspaper to dry.)

Human beings, I've come to realise, have no place in this particular system, short of providing food or shelter. (Even then, there's no certainty about the consequences of our providing food. Evidence suggests that it's beneficial but it may affect the distribution of some species, making them dependent on human feeding.) Our influence is malign in virtually every way, including those we don't yet know about or understand. But within the limited framework of the artificial spaces of nature we have created, learning to stand back is all we can do. We cannot alter the balance of numbers or of life, nor should we. When we try, we damage. Trapping magpies and killing them doesn't reduce the number of lost 'songbirds' – all it does is reduce the number of magpies temporarily, and turn people who purport to love birds into violent and possibly lawbreaking bird-killers. Shooing hawks away may protect our doves or garden birds for a few minutes but it does nothing more because our lives do not depend on our so doing and the hawks' life does. It shouldn't do more because hawks are beautiful and necessary, as is every other naturally occurring species. For garden birds, we may provide the safest places we can for roosting, nesting and feeding. We may offer a regular supply of food, keep their environment clean, recognise each one we can, name it, listen to it, appreciate it, scare cats away from it but after that, everything is in the fickle, mysterious lap of those unknown gods who attend to such things – fate, or nature or whatever one wants to call the imponderable forces behind this strangely ordered world.

The Unlikely Cupid

From under the overhanging, soft-haired leaves of the wet *Alchemilla mollis*, I pick up the little rectangular stone trough which has over the years filled up with small pebbles of assorted shades and sizes, gathered up on walks. I tip them out onto the path as I turn the trough over and there's a sudden scatter of assorted-sized woodlice, a swift fourteen-legged rush into the stones. On the under-surface of the trough, two creatures are making their measured way in the unexpected light. They are *Deroceras reticulatum*, grey garden slugs, one large, one very small. The larger moves softly over the surface of the rough stone and turns its head and antennae with a leisurely, measured grace. Slow and glistening, its pale body is a translucent grey, pale violet and cream. A bubble of eggs clings to the corner of the trough. If I knew nothing of these creatures, what would I think? What might I imagine if I didn't know the word 'slug'? But I do know the word slug, and most of what I know of slugs has to do with opprobrium and disgust, with killing and removal. I know too about slime, and the awful depredations that a slug may do with his implacable, devouring jaws, his rows of tooth-like radulae, advancing on his soft and moving foot. To look at slugs, you wouldn't really

206

think that they could be so reviled, could arouse so much anger, inspire the degree of blood- or possibly slime-lust that they do, these members of the second largest phylum of the earth's creatures, the Mollusca, these belly-creeping gastropods, gliding serenely through our gardens on their own personal trail of mucus.

Watching this stately progression, I think of the American poet and environmentalist Gary Snyder's comment: 'Life is not just diurnal and a property of large interesting vertebrates, it is also nocturnal, anaerobic, cannibalistic, microscopic, digestive, fermentative, cooking away in the warm dark.' Or indeed, in the cold dark, and the cold light.

The slug is an odd, unlikely creature. I try to look at it from a distance, from the view of first encounter. I try to dismiss from my mind every prejudice that comes with the word 'slug', the knowledge that there are few words in English less gracious, more co-opted as insult, containing as it does most of the letters of the word 'ugly' as well as suggestions of laziness, reluctance, sloth. Is there any word that demands our disgust, our contempt and disregard, more than slug? I look at this alien form moving strangely majestically, nakedly, its glistening body alarmingly open and vulnerable. I replace the small trough, scoring out an indentation in the stones before I do it so that my fellow sharers of the garden may resume their dark and quiet lives.

Slugs may not – in fact, do not – command the same degree of interest or affection as some vertebrate species, or any interest or affection at all but indeed the reverse. By suggesting that they might be worthy of even a small amount of attention is not intended to suggest that I feel what one American writer, commenting on 'nature writing', called scathingly 'mystical oneness' because I feel no particular mystical oneness with slugs, or much else for that matter but because I recognise that I might have underestimated, or simply never have known or appreciated their full, useful virtues and I'd like to make what effort I can to consider even their small, decried place in the world. (I'm not, it appears, alone. I have just read an article about Giant African land snails being kept as pets. I find the article encouraging. It's yet another indication that someone, somewhere is willing to extend their interest and care, and who knows, even their affection to the unlikeliest of subjects. The problems associated with keeping Giant African land snails do seem singular and interesting, though, dealing with the vexatious matter of excessive slime being only one.)

Not only used as a term of contempt, the reductive word 'slug' itself seems to ignore the place of these creatures in the grand of panoply of molluscan life. Of at least 50,000 molluscs (and possibly as many as 200,000), there are 13,000 named genera and some 62,000 described gastropod species. The systematics – the scientific naming and placing of these creatures – has recently undergone a massive process of change. In 2005, a revision of the entire system was published in the journal *Malacologia* in a paper written by two scientists from the Natural History Museum in Paris, Philippe Bouchet and Jean-Pierre Rocroi, which described the reclassification of gastropods according to phylogenetic principles which take evolutionary relationships into account. According to the new naming, the majority of land snails and slugs belong to the 'informal subsection' of Sigmurethra, in the clade Stylommatophora.

Sometimes it seems that nomenclature is, if not all, then a great deal. Give a man, bird, beast or slug its Linnaean name and instantly it is elevated from being merely what society deems it to be – in the case of slugs, not very much – to join a company of the designated, the named and chosen and if by virtue of Swedish thoroughness we all belong to this particular company, it doesn't diminish the fact that *Deroceras reticulatum*, *Limax flavus* or *Limax maximus*, *Arion ater* or *Arion hortensis* sound distinctly more distinguished than 'slug'. Imperial, in fact.

It may be the fault of the disgracefully biased old rhyme about the differing qualities of male and female children, or their being of the same class of creature, but slugs and snails often occur together in the mind although snails seem to be generally regarded as a cut above slugs. They, at least, aren't naked. They have decently covered their semi-translucent, almost gelatinous appearance with decorative and useful shells, shells which bestow a characterful edge sufficient to engender cartoons and allow their portrayal as toys and garden statuary and to become internationally famous as Brian, the snail from *The Magic Roundabout*.

Slugs and snails, as everything else, have their place in the scheme of life, in the food chain, in the ecology of the earth; a purpose, you might call it, even if it's a purpose that doesn't always accord with our own. They eat things, and are eaten by other things and they break down organic matter. They eat rotting vegetation, dead leaves, fungi and, alas, what we grow in our gardens. We put them in an impossible situation. We create gardens, fertilise and improve the soil, grow the lushest, finest plants we can, in the kinds of arrangements that best suit slugs, with dense plantings, dark places, shrubberies. To make the life of a mollusc even more trying, we do it all in a climate most conducive to encouraging them to thrive, a climate both temperate and frequently wet, ideal for a creature with a body that requires to be permanently moist. We are orderly and tidy and remove the dead and rotting stems and leaves and vegetable matter on which slugs feed. To defend our plants, we often compromise the natural environment of our gardens under a hail of livid blue poison. We put down slug pellets containing methiocarb and metaldehyde, both lethal to creatures other than slugs, and proven contaminants of water supplies. Slugs and snails are someone's bane but someone's food, eaten by birds (thrushes are particularly fond of snails), by ducks, frogs, toads and some beetles, all of which can be affected by eating poisoned slugs and snails. I don't know if my doves eat slugs, but I lift out any I find who have crept into the doo'cot and return them to the stones outside. They may well return to the hostas and surmount, by means of gastropod artfulness and slime, the ash barrier I have extended round the plants, as well as the few rotted leaves from last autumn I have left designed to attract them before they reach the new fresh plants, and probably they will succeed in nibbling holes in the leaves and if they do, they will have eaten well. Who am I to deny them? (I know though that I wouldn't take this sanguine view if I were dependent for food on the plants they ravage.)

Slugs reuse slime trails, travelling on their own pre-slimed routes, rather like rail commuters, a manoeuvre that must be the ultimate energy saver. By following their own trails, they find their way back to where they came from, demonstrating 'site fidelity', that impulse that stirs in so many of us, the impulse to return to the place we regard as home.

Mostly hermaphrodites – possessors of both male and female sex organs – slugs and snails are able to mate with any other member of the same species, and some are able to self-fertilise. Their mating behaviour can be both spectacular and strange – some species engage in elaborate mating that involves twisting their bodies into a single intertwined rope, while the beautiful, spotted leopard slug, *Limax maximus*, mates at the end of a long string of self-created mucus.

Oddly enough, the origins of Graeco-Roman symbolic portrayals of romantic love may perhaps be traced to the mating behaviour of Stylommatophora. In one of the most remarkable phenomena in the natural world, some species of snails thrust a dart or 'gypsobelum' (described even in scientific papers as a 'love dart') into their prospective partner before mating. The function of the dart has long been questioned although Dr Ronald Chase of McGill University suggests that it transfers a hormonal secretion, which, in affecting the sperm-storing tract of the recipient, increases the likelihood of the sperm success of the donor. Composed of calcium carbonate, chitin or cartilege, the darts vary between species. Bladed or veined, spiralled or straight, dagger-like or curved, barbed or smooth, all appear to fulfil the same purpose. Dr Chase suggests that since one of the species employing the 'love dart', *Helix aspersa*, is common in Greece, and the ancient Greeks were able observers of natural history, the origins of the myths might be explained. Can one ever look the same way at a slug or snail again, knowing that portrayals of Cupid and Eros with their waiting bows and arrows symbolising the pains and pleasures of love may be traced back to these unlikely symbols of erotic passion?

In a garden centre I find a book, a jolly little book, one actually intended to be funny, which is composed of suggestions on how to kill slugs. With its pointed sticks and hot pokers, drownings and burnings, it reads like a volume of torture and wanton killing, a kind of mollusc *Gesta Francorum* for our time. On any gardening programme, a major topic of question and discussion will be that of the removal or killing of slugs. More than once I have heard the suggestion that they should be impaled on pins, and while so little is known about how much slugs' neural or sensory equipment allows them to experience such an outrage, an outrage it is.

In a study carried out at the University of Oslo in 2009 for the Norwegian Scientific Committee for Food Safety, the brains and central nervous systems of a variety of invertebrates were examined (not with a view to their being safe to eat, but for the purpose of establishing how they should be handled and whether or not they are sentient and may feel pain. Incidentally, information I find suggests that eating raw slugs can cause meningitis. I take note. There is no indication of what eating cooked ones does).

Among the animals studied were crabs, lobsters, octopuses, worms, snails, slugs and spiders. The central nervous system of each is described, and their capacity for sentience and feeling assessed. While suggesting that the gastropod central nervous system is regarded as one of the most challenging problems in current comparative anatomy, the author describes the gastropod system of paired ganglia (the collection of nerve cells controlling a particular action such as locomotion), neurones and glial cells (the names, the terms, the cells, indeed the functions are no different from the ones within our own central nervous systems). Snails and slugs, devoid of the capacity for sound and, largely, for sight, are entirely dependent on a variety of sense organs for their interaction with an external world. Possessing statocysts, organs of balance that allow them to detect gravity, and proprioceptive sense organs that give them the ability to judge their own position in relation to objects, they are dependent on these chemoreceptors and mechano-receptors to negotiate their way in a complicated environment, to find food and mates and to defend themselves. They are capable of simple learning, called habituation, and display associative, spatial and social learning. They perform appropriately on preference tests and have been shown to have short- and long-term memory. The author of the paper implies that we might be wrong in our assumptions about the limited capacity of invertebrates to suffer. Suffering, he suggests, is a private experience impossible to measure, and the responses of invertebrates to harm are often similar to those of vertebrates, responses that might indicate 'a level of consciousness or suffering that is not normally attributed to invertebrates'. Is it untoward or ridiculous to consider the possibility of consciousness or intelligence in a slug, a snail, a worm? What do we know about the brain capacity of slugs? Do we wish to know that we live in a world of sensibility, to extend the borders of our pain, to forget that it is thought acceptable to kill a living creature with pins?

213

In 1881, Charles Darwin published the last of his books. The result of prolonged and astonishingly comprehensive observations, he called it: *The Formation of Vegetable Mould Through the Action of Worms with Observations on Their Habits* (now more often abbreviated to the snappier title of 'The Worm Monograph') and in the introduction, he writes:

> As I was led to keep in my study during many months worms in pots filled with earth, I became interested in them, and wished to learn how they acted consciously, and how much mental power they displayed. I was the more desirous to learn something on this head, as few observations of this kind have been made, as far as I know, on animals so low on the scale of organisation and so poorly provided with sense-organs, as are earth worms.

(Worms, of course, have an easier time, being correctly regarded as entirely beneficial. They do suffer, however, from the same aesthetic judgements as those applied to slugs and snails.)

In Chapter One, 'The Habits of Worms', under the heading 'Mental Qualities' Darwin assesses aspects of their behaviour: their timidity, their enjoyment of food and their sexual urges – all sufficiently powerful to encourage them to overcome their fear of light – and the possibility of their having social feelings. Musing on the question of intelligence, he suggests that deficiency in sense organs need not necessarily mean the absence of intelligence and indeed, says that 'when their attention is engaged, they neglect impressions to which they would have attended; and the attention indicates the presence of a mind of some kind'.

In the next chapter, having closely observed the ways in which worms stop up the mouths of their burrows (a chapter that includes some astonishing insights into the behaviour, strength and tenacity of worms), Darwin compares the methods which a man might use in undertaking the task of plugging a cylindrical hole with leaves – 'the guide in his case would be intelligence' – with those of a worm. Worms act not just from instinct because instinct wouldn't tell them how to deal with types of leaves unfamiliar to their progenitors, and since they organise their leaf-collecting in a way that precludes chance, 'intelligence alone is left'. In his conclusion, Darwin writes that, while it may seem improbable that worms possess intelligence, too little is known of the nervous system of worms for the possibility to be discounted.

'With respect to the small size of the cerebral ganglia, we should remember what a mass of inherited knowledge, with some power of adapting means to an end, is crowded into the brain of a worker-ant.'

Should the potential presence of intelligence, feeling or consciousness of any degree by itself determine the way we act towards other creatures? Which creatures, and in which ways? What degree of sensation? (Again, I remember the Jains and the microbial life around and inside and all over us.) How is it possible not to trivialise the arguments by detracting from large harms by concentrating on possibly smaller ones?

As well as the consideration of the effect on creatures, it's for ourselves that I worry, the outrage to ourselves of sticking pins in anything animate which appears to me to do no good for the sum of human sensibility. The English philosopher John Locke in his 1692 treatise 'Some Thoughts Concerning Education' (*Some* thoughts? Every thought! Everything ever considered by anyone, anywhere on the subject of the raising of children, their lives and education, it may be the most precisely detailed, comprehensive, exhaustive of any document ever written on earth, apart from Darwin's Worm Monograph) discusses children and their treatment of animals, observing that children should be discouraged from 'ill using' creatures because 'the killing of beasts, will, by degrees, harden their minds even towards men'.

More and more I have the feeling that, read or listened to in the future, books and programmes that tell you how to kill slugs with pins or fire will evince the wincing shame with which we now listen to old recordings and read old texts of every sort that demonstrate the falseness, the damaging, casual cruelty with which humanity has treated its own and others.

'I think,' Locke wrote, 'that people should be accustomed, from their cradles, to be tender to all sensible creatures, and to spoil or waste nothing at all.'

As if to redress the balance in favour of respect for worms, Darwin calls Chapter Four of the monograph: 'The Part which Worms have Played in the Burial of Ancient Buildings', and reminds archaeologists of their considerable debt to worms. He reports on measurements, calculations and examinations of every minute process of burial. He discusses materials, methods and depth of burial. (He does seem to have pressed his sons into action on this one:

When my son William examined the place on January 5th 1872, he found that the pavement in the three holes lay at depths of 6.75, 10 and 11.5 inches beneath the surrounding turf-covered surface [. . .] My sons Francis and Henry visited the place in November 1877 for the sake of ascertaining what part worms may have played in the burial of these extensive remains.

Dutiful chaps.)

Darwin's conclusion is: 'The cases given in this chapter show that the worms have played a considerable part in the burial and concealment of several Roman and other old buildings in England . . .'

Perhaps our approach to Mollusca, Annelida, Arthropoda, Nematoda and all of the other small, disparaged and necessary creatures of the earth should differ. Perhaps we should try to find some ground for mutual cooperation. Some years ago, an Australian naturalist discovered that the answer to the problem of the mould and fungus that had developed in the bathroom in his damp, low-lying Sydney house was to introduce a number of the mould-eating slugs that he gathered from his garden. After assembling his work team, he set them loose in his bathroom, providing them with a small ceramic pot in which to make a home. (The pot, he notes, one previously used for the burning of incense, had a pattern of cut-out stars and moons.) Every night, the slugs crept from their celestial ceramic home to feed on the mould, cleaning it with supreme efficiency before, every morning, creeping back with their innate instinct to return to their home. The only time they failed to turn up for work was during the breeding season, when they escaped down the drainpipe to seek romance. The situation sounds ideal. While engaged in the business of bathroom-cleaning, slugs would not be eating garden plants. Knowing this has made me anxious to investigate the possibilities. Do I detect a business opportunity, a slug-shaped gap in the market? (There is a handy mould-eating slug in Britain – *Limax flavus*.) A slug-staffed cleaning company. There would be no need to rush around tidying the day before the unfortunate who cleans up after you arrives, so that they don't have to witness the morbid desuetude of your existence. You wouldn't need to lurk guiltily as someone else sorts out the week's horrid detritus. The company name might be difficult, but the strapline wouldn't: 'Conscience-free cleaning because slugs don't care'.

EARLY SUMMER

May 23rd

Late May, and we walk out to the road above the harbour, beyond the streets of fish-houses, past the rows of allotments facing the sea. It's probably not the best day to walk because in spite of fleeting, cloud-blown sunshine, a ferocious wind's blowing. By now, the allotments are all netted and neat – lines of raspberries and peas are caned and trussed, beds of strawberries and roses are growing towards high summer. Behind us, the harbour's busy and the river too, the narrow channels that open today onto the wide sea are lined up with huge oil supply ships heading out in relays towards rigs far beyond the horizon, beyond Shetland, towards Norway. It's all we can do to walk at all, the wind is so fierce.

Early summer – but suddenly, these strong winds, this undertow of chill. Yesterday there were forecasts of hill frost and today, photos in the newspapers of people skiing in the Cairngorms, children frolicking in snow.

The winds have been blowing for days but today they are no longer just strong winds, the ones that have caused the streets to be carpeted with leaves of sycamore and lime. Today they are gale force, 80 mph, the kind to tear off roofs, to toss the largest of the ships as they ease beyond the navigation channel of the harbour towards the open grey sea, to tilt them into the waves at vertiginous angles. In the distance of the water, their stern grey edges are softened by the vaporous, rainbow haze of

221

blown sea. On days like this, ordinary walkers could be whisked from cliffs, lifted like Thumbelina, feet over head, high over the masts and spires and roofs. When we try to speak, our words are snatched from our mouths before they're spoken. Within moments, our eyes are stung by salt, ground red by grit and sand. Behind us, the city is busy on an ordinary Monday afternoon. Stopping on the road and looking back, I can see the shop signs, the clocks, the spires, but since there aren't any trees to show the strength of the wind, the city looks solid, implacable and quite unmoved. It's only the sea that shows the power of the storm.

We have to bend, our heads down against the weight of air. It slips roaring into our ears and round our heads. Deaf and mouthing, we take the lower path, one that leads nearer to the water, almost but not quite below the line of the wind, but the moment we move onto it, it is suddenly quieter. We're still wind-boxing as we walk towards the breakwater, trying to hold the binoculars steady enough to see the eiders and herring gulls massed in the calm, protected space. A few cormorants stand airing their wings, holding them high like small black dinosaurs or armorial dragons. A juvenile stands on a rock, untroubled by the gale, within the protecting bounds of the breakwater. A coterie of cormorants bobs on the swelling water, serenely, like ladies by a bathing machine. There are few of the birds people come here to watch: the migrants, the seabirds, rare and common, the gulls, lesser and greater black-backed, the yellow-legged, the glaucous or Iceland. There are no skuas, shearwaters, terns, scoters or mergansers. It's sunny but then the rain begins, a fine-blown net glittering with sunlight, what we always call a monkey's wedding.

In spite of the wind, there is a constant passage of vessels going off towards the oilfields, bucking and swaying until they're almost out of sight.

Beyond the harbour, a pale, wide, deep-breathing sea, the sand curving for miles like an arm around the bay. There's a thick patch of mussel and cockleshells in between the rocks, dropped by feeding gulls. A sign suggests that it might be best not to swim in the sea (who would, here, even on a day of perfect stillness? Who would, anyway, with warning signs about sewage pollution, with the almost invariable, invincible cold?). We walk on, holding on to the railing with both hands, trying to breathe although there's very little opportunity for such a luxury when you're fighting with the wind. By Girdleness, round the headland, we can hardly stand. We turn our backs towards the wind. In the harbour, a seal dips and surfaces. There are no dolphins today, no harbour porpoises. Further out, there are no humpback or minke whales (a few weeks after this, a veteran cetacean-watcher will witness the birth of a bottlenose dolphin in the bay). Although soon, it'll be the peak season for all the cetacean species: the white-beaked dolphins and rare Risso's dolphins too.

Facing the wind, we force our way back, leaning upright against the air. We laugh at the way we're forced to move, tottering forwards, driven on, and when we get into the car, there is sudden, lovely silence and stillness and it feels as if we've been in the gale for longer than we have.

We drive back through the city over ripped-down leaves and layers of new, fresh green on the roads, through arms of broken branches

223

which extend jagged edges of white raw wood like shattered bones of hands towards us. Passing Duthie Park, we see that even the pond has turned tidal, raised into silvered streamers of froth and waves. On a back road, a scaffold of fallen trees forces us and the few cars behind us to turn in formation, all of us patient and neat like dancers in a chorus line, three, four points and back the way we came. In a suburb along the Deeside road, a roof has been lifted neatly from a building and fabric, curtains perhaps, or insulation, flaps eerily from the space where joist and slate used to be. In a street nearby, a car's buried under leaves, the windscreen caved in by the weight of fallen trunk.

Among the torn branches, there will be nests and newly hatched birds. It has happened at the worst time. I don't know what this wind is called. There will be one, a name for a gale-force wind that summons maximum strength just when you thought that if it wasn't quite summer, then it was definitely spring.

And now, for months, all summer and into autumn, torn branches and half-trees will lie around the city, leaves alternately drying and being soaked by rains. Eventually, months on, jagged stumps will be dug up and hauled away, leaving roundels of trunk and sawdust, tree-less spaces, squares of emptiness, new vistas open to the sky.

May 28th

This morning as usual I check Aurora Watch. The website has been updated. The appearance has changed slightly; the words with which the disappointing news is conveyed. Now, instead of 'No major activity detected', it says, 'No significant activity' (leading me to reflect that, in all the time of watching, this change of wording is the only auroral activity I've identified). There's a qualitative difference in the use of words. The updated one feels less inclusive, less kindly. The suggestion that it might have been merely a failure to detect encouraged me. The second usage is final. It is a judgement.

June 3rd

In the doo'cot this morning, I find that something has reduced one of the doves to a neatly stripped skeleton and a few wing feathers. It's something I've never seen before. I've seen what cats and hawks can do to doves. Cat attacks are the worst. It's just mauling, all blood and little eating. Hawks dine efficiently but don't leave skeletons. (They leave lots of feathers, selected entrails and feet.) Besides, this was inside the doo'-cot. The only means of entry would have been through some of the small gaps in the sides of the wire of their door. No cat or hawk could have got in this way. I consider rats but they too are messy eaters. What else could have got in?

Thinking over the possibilities, the most likely is a mustelid – weasel, stoat or mink – none of whom I have ever seen in the city, never mind the garden. A mink would be too big. I drove past the corpse of a mink on the Deeside road one morning, a heap of glossed fur gleaming in the sun. (There are fewer now in this area than there used to be, since the

225

mink eradication programme at the River Ythan.) The only time I see stoats or weasels is as a streak of slick movement by the side of country roads, or as twists of discarded, traffic-flattened fur. Even then, I'm not sure that on meeting one, I'd know the difference. (*Weasels*, Dave would say, *are weasily recognised. Stoats are stoatally different . . .*) I examine the area outside the doo'cot, then the garden and the lane but find nothing: no scats, no footprints, no other evidence at all. I imagine whatever small creature it was, having dined so amply and so neatly, engaging afterwards in a joyous exercise of parkour, that sport of running, scaling, jumping; a ninja-style, wild flight over the neat back walls and gardens and off to the open country not far away.

June 7th

A golfer who killed a fox with a golf club while playing at a course on the outskirts of the city recently met a kind of justice. The fox was tame and well known to the golfers but was, the man said in his defence, paying unwanted attention to his golf bag, which was why he struck and killed it. The miscreant was prosecuted under the Wild Mammals (Protection) Act 1996 (Scotland), found guilty and fined. The fox, the man said, 'looked at him in a threatening way'.

By chance not long ago, I read a letter to a Glasgow newspaper in which the writer complained of a brief encounter with an urban fox. He described the look the fox directed towards him as 'sullen'. Interpretation of fox expressions seems to be a flourishing art.

The misdemeanours of foxes are an ever-popular subject, one that seems to have grown in fervour since the report of one having entered a house in London and bitten two young children. After this, stories about urban foxes began to abound in newspapers, even in areas like this where, as a result of there being very few foxes, there's little to report. But even an absence of foxes hasn't prevented a proliferation of urban fox stories and reported sightings. (I don't have either. I did hear a rumour of a single fox being seen in the neighbourhood but although I've looked, I've neither seen it nor heard of it again since. I have seen one dead fox on the road outside the city.) While it may seem remarkable that a fox would prefer central London to here – only because central London would seem to offer fewer possibilities for what might be deemed a natural fox diet and even more dangers – foxes in this city are comparatively rare. I'm not sure why but it could be because foxes have no need to be in a town when the countryside's so near. It's uncertain why there has been an increase in urban fox populations – it may have happened because numbers in the country have increased, reducing territorial prospects, or they may have become adapted to urban living and to increasing levels of waste. In a book about the wildlife of Aberdeen written thirty years ago, the author comments that foxes remained creatures of open ground, having not yet attained the urban habits of London foxes. Might foxes have 'site fidelity'? Might they have attachment to places now built on? In urban areas where they do proliferate, might they simply have been there first?

After the trial of the fox-killing golfer, the local newspaper reports in a generally febrile tone that there were 17 fox sightings in the city last year. A councillor interviewed on the subject suggests that the foxes be trapped and removed to the countryside on the grounds that 'city folk aren't used to them'. He is worried about the spread of disease because foxes carry fleas and ticks. (A bit like dogs, in fact.) A lady in the centre of the city has seen the same fox twice. 'It was three foot!' she's reported as having said but whether this was height, length or girth is, alas, not recorded. There's so little to report on the matter that every fox fact, every fox incident from recent years, if it happened within 500 miles of here is added to the article, including the sorry account of an unfortunate who, happening to fall asleep in a Midlothian graveyard (over a hundred miles away), lost both nose and fingers in a 'suspected fox attack', illustrating, if nothing else, the folly of falling asleep in a Scottish graveyard. (The certainty being that if a fox doesn't get your extremities, the weather will.) Besides, having suffered such an experience, anyone's wildlife identification skills might be less than wholly accurate.

June 15th

The swifts are shrieking overhead as I work in the garden, as I hang out washing, clean the doo'cot. Even on the grey days, these small, miraculous birds are here. Most are swifts but there are house martins and swallows too. Swifts, *Apus apus* of the family Apodidae, swallows *Hirundo rustica* and house martins, *Delichon urbica*, both Hirundinidae, live lives of almost perpetual flight. Their speed, the ephemeral nature of their presence, the way they appear, disappear and conduct their lives according to the mysterious impulses of migration make it a life most difficult to understand.

Swifts copulate on the wing. They drink by skimming the surface of water. The American ornithologist the late Louis J. Halle describes them as being possessed of feet that cannot do more than cling to vertical surfaces and wings too long to fold, as the wings of other birds do, into their bodies. They inhabit, he suggests, a world where humans are marginal, one belonging only to birds and insects. Nevertheless, having adapted themselves to our civilisation, they have been among the most abundant of our city birds, but swift numbers are falling. They've declined by 26 per cent since 1994 and now they're 'amber'-status birds, threatened and reducing for the same reasons that so many species are reducing in number: the effects of pesticides, loss of habitat, shortage of food. The way we build has a direct effect on swifts and their ability to create nest sites. We modernise our roof tiles and gutters, sealing every space and gap. Our methods and materials keep warmth in but swifts out.

In *Notes from Walnut Tree Farm*, Roger Deakin recounts a dinner-party conversation he has with two fellow guests. One complains that green woodpeckers have pecked holes in one of her clapboard walls in order to nest. She tells him of the cost of repairing the damage, of how she has hung balloons around the eaves to deter the birds. When he suggests that it might be rewarding to have a nest box built for the birds, the other guest complains of the swallows nesting under the eaves of her house to whom she objects because of the mess and the noise.

229

Her husband, she says, has knocked down the nest and installed wire to keep the birds away.

June 20th

I'm talking to a friend who is about to go home to Latvia for a holiday. The temperatures there are breaking records. She tells me that the cherries in Latvia have ripened early. Every year when she is at home to celebrate her name day at the end of July, her mother makes her a special dessert with the first cherries. This year, her mother has promised a dessert of cherries and strawberries. 'Both together!' she says. 'What's happening?'

What is happening? Again, I regret my failure to record natural events, first arrivals, final sightings, migrations, first falls of snow and frequency of rain. I didn't because I wasn't far-sighted enough or sufficiently aware of the fragility of everything around me. Perhaps I should have known that the year we moved here was significant, possibly the final time that the atmosphere of earth would be as it was, possibly at a moment beyond recall, when the balance of gases in the atmosphere would pass beyond the limits of their safety.

I look again at the photograph of Bec in the garden but there's nothing in it that could serve as an indication or a warning. Just by looking, you can't know anything of air. There's nothing to indicate, nothing that warns. You can't see change in air, or sniff it on the wind. How do you know? Because there seems to be more rain, a different kind of heat?

230

It has been known for a long time that the earth's climate could be subject to alterations over long periods, and while there has always been speculation about possible causes of variations in temperature and rainfall, it was with the vast and sudden increase in human industrial activity that suspicion began about its possible repercussions for the climate. In the early years of the nineteenth century, the French physicist Joseph Fourier suggested that earth's atmosphere might act as a greenhouse, trapping heat, and when later that century Svante August Arrhenius, a Swedish physicist, began to take an interest in the relationship between CO_2 levels in the atmosphere and temperature, he related possible future rises in CO_2 and possible consequent rises in temperature to the burning of coal (although he considered it to be a generally desirable prospect as it might produce a more clement climate – the man was Swedish and might have been writing in January). Whatever the consequences, he believed that they wouldn't happen for a long time. Other scientists, often from their interest in the timing and nature of ice ages, began to study atmospheric change and the influences of human activity on climate, concluding eventually that observable changes in temperature might be anthropogenic, and that what was happening to the climate was happening fast.

On his comprehensive, startling website, the American physicist Spencer R. Weart, author of *The Discovery of Global Warming*, explains in precise detail the history of the unfolding of knowledge and research into the progress and processes of changes in climate, and illustrates each step with references to the scientific data. As gripping and as horrifying as any work of fiction, the scientific, political and social complexities are documented and analysed over years, from the first studies of the effects of CO_2 until the present day. (Throughout these accounts of the growth of knowledge there was, until remarkably recently, a feeling that there was no urgency, no sense of immediacy. Change takes a long time to happen – besides, how could humanity have influence over the vastness of the planet?)

We live today, as we have done for the past 10,000 years, in the geological epoch called the Holocene. For most of its duration, temperatures have been stable within narrow margins – the composition of gases in the earth's atmosphere, too – safe and amenable, the climate sufficiently constant and indeed mild, to allow life on earth to flourish. But at some moment everything changed and the changes, frozen into time, have been revealed by drilling through millennia worth of ice, more reliable in information than anything written in any human document. As techniques of ice-drilling improved during the latter half of the twentieth century, what was discovered by measuring oxygen isotopes in ice cores from the end of the last Ice Age, 11,000 years ago, was startling – changes in atmospheric compositions were possible in a much shorter time than had ever been believed. There are geologists who now refer to the present geological epoch as the 'Anthropocene', a term first used by the Dutch Nobel Prize-winning chemist Paul Crutzen, who believes that a new epoch began in the late eighteenth century with the beginning of industrialisation and, more specifically, with the invention and use of James Watt's steam engine in 1784. (While the consequences of human action have been and will be damaging and profound, Professor Crutzen suggests that things might have been a lot worse had there been a different set of chemical reactions involving halogen and chlorine in the atmosphere, a situation that would have caused catastrophic change but which didn't happen through luck. Good to know.)

For most of human history, until 200 years ago, concentrations of carbon dioxide, a necessary component of the atmosphere of the earth, were 275 parts per million (ppm). In 1987, they were 350 ppm. In the years since then, they've increased to 394.97 ppm. In these figures, it seems, lies the history of life on earth and its future.

Another scientist who agrees with the idea of the Anthropocene is Dr Jan Zalasiewicz of Leicester University, who puts the date of the major changes, ones that will be recorded in future rock compositions, at 1945, the beginning of the nuclear age. The period after the Second World War was also the beginning of an unprecedented rise in the human population and in economic expansion. Mass travel, mass production, invention and expectation are, certainly for the Western world, difficult habits to relinquish.

In 'Planet Zoo', an article published in 2010 in the online newspaper *The Morning News*, the American writer Anthony Doerr presents the current situation as a metaphor – we're all driving down a foggy road, past signs warning of a cliff ahead. We can stop, slow down or 'put the kids in the backseat to work sewing parachutes', but what we're more likely to do is to minimise the cliff, think it isn't really there, hope it's further away than we think, that we can stop at the last moment, or that the kids are particularly good at making parachutes. In a funny, analytical piece, Doerr examines the earth's predicament, feeling himself, as we all may, no less implicated in its fate than those who act with intention or wilful blindness. He writes of his appreciation of the benefits of the twenty-first century and considers future possibilities for dealing with climate change such as geoengineering, those way-out-sounding proposals for mechanisms to reduce solar radiation or to remove carbon from the atmosphere. Quoting Christopher Marlowe's *Dr Faustus*: 'The god thou servest, is thine own appetite', Doerr asks, 'Was he wrong?'

From now on, we'll look back and wonder how it happened. I try to remember the moment when I first knew about it, and if there was such a moment. When might it have been possible for cessation or change? I ask myself constantly about my part and my responsibility. We may, through lack of alternatives, all have contributed. Had we recognised the moment when we might have done things differently – the minute, the year, the decade – what might we, as individuals, have done? What more than the things we try to do, to be aware, perhaps; use less, consider more? Then, as now, we may have had too few personal or financial resources to live differently, too little information, or inclination, too little idea of what might happen, too much desire to believe it not to be true, too much fear that our lives might change. Most of us, after all, didn't promulgate the idea that, among other damaging things, to drive faster, larger cars would invariably change us from the uncertain, not-always-beautiful creatures that we – almost without exception – are, into figures of improbable glamour. That some of us tried, and continue to try, the experiment and fail to perceive that we remain much the same shows more susceptibility than evil. Everything in our way of life is predicated on the idea of the necessity for progress and on what may be produced, and if all of it seems desirable and irresistible, that is as it was designed to be.

In the *Tao Te Ching*, Lao Tzu writes his thoughts on the perfect state, a state that might have machines but wouldn't use them, might have boats and carriages but wouldn't use them either (although he doesn't explain why they might bother having them in the first place), a state that would use knotted cords for record-keeping and:

Though the neighbouring states are within sight
And their cocks' crowing and dogs' barking within hearing
The people will not go abroad all their lives.

(The early Taoists were well known for exaggerating for effect. Still, I kind of know what he means.)

Everything may change but we don't know how. Sometimes it seems too much to deal with. The facts and the numbers are too vast, too unknowable, and all of us just too small.

'See my works, how fine and excellent they are!' it says in Ecclesiastes, and the author, whether Kohelet the Preacher or King Solomon, was seldom wrong. 'All that I created, I created for you. Reflect on this and do not corrupt or desolate my world, for if you do, there will be no one to repair it after you.'

235

June 21st

The day of the Summer Solstice. The moment itself is at sixteen minutes past three on a damp Tuesday afternoon, when earth's axial tilt is most inclined towards the sun. In an attempt to be latitude-neutral, I remember that it's the day of the Winter Solstice in the Southern Hemisphere.

June 30th

A single tiny fly is hovering persistently around the kitchen sink. It's too small to be moved, too delicate to touch. If I turn on the tap, it'll perish in the almost unseen spray.

I think of the radio programme I heard a few months ago on the subject of belief, in which a prominent Jain was being interviewed. The interviewer was trenchant and enquiring and wouldn't let questions go by easily. Again and again, she asked about how far the life and practice of the Jain can deal with the knowledge of the unseen and the microscopic. The interviewee seemed helpless before the questions. He had clearly thought of it all before but had no answers.

'More research is needed,' he said but knew as we did that he was facing the impossible, the ultimate paradox, that of our being and of how to be, that of invisibility and power, the infinite interrelationship between one life form and another, most of which, as Eric Chivian says, we know nothing about. The prominent Jain knew as we did that his peaceable, gentle religion grew and spread its ideas before microscopy, before we knew how much exists that we can't see, or about the connection between life forms, so many of which we still don't understand.

July 11th

This morning, I have a date with time. I walk the short distance from the house in heavy summer rain to the place where I'm to meet the others who'll be there too, to gaze and reflect on a place as strange and unlikely as any you'll find hidden in the centre of any city. I've often done this, walked the few hundred metres along the main road that leads four lanes of traffic between the western suburbs and the centre of town, stopping by the steep bank where the trees begin. I've levered myself up onto the retaining wall and scuttled up the slope of trees and broom and shrubs to stand like a prisoner looking though a wire fence to the other side. I've driven past it hundreds of times, by now hardly even noticing the large yellow warning signs spaced along the perimeter, the ones that say 'DANGER!' I've kept out because the high chain-link fence has obliged me to. I've thought about what it would be like to find a way in, to climb over, or break through the fence past the warning signs, to take up a watching position on the other side and sit, looking, thinking of the life of the place, today and backwards in time, of the nature of water and of stone. I've never done it because I've always been afraid to, especially on my own because the edges are so steep. Over time, the rim of land has grown narrower, leaving less and less ground on which to stand or to walk. (There's no place now for the roe deer who were said to live behind the fence. They must have escaped, or drowned.)

What there is behind the chain-link fence is a quarry although the word 'quarry' itself feels wrong – 'quarry' seems to describe a dry and bleached-out place, spent and full of stones, all rutted roads and lorries churning dust. This one must have been like that once but now is all depth and silence. When I saw it first, years ago, not long after we came here, it was already half full, but over the years the water has risen. On days of sun, it looks benign. It could be a forgotten water-park or an abandoned boating lake but with a post-nuclear air; people-less, boat-less, only a few ducks floating in the quietness, just rock surrounding a body of still, apparently shallow water. In winter, in snow and near darkness, it's like a single, giant obsidian eye, staring at the sky.

You could live here for a long time and have no idea what you're passing because there's nothing to be seen from the road, only the warning signs to indicate there's something behind the bank of earth and trees. I'm used to the thought of it now, of the almost uncanny uniqueness of the place but I began to think differently of it when, on a day in winter a couple of years ago as I was passing, I saw a new sign. It was just a message on a board, the unassuming, ordinary kind used by estate agents to advertise commercial property, the kind that becomes part of the landscape of things no one wants to buy, which they drive past day after day until at last, someone gives up and takes it down. 'For Sale', it read. Stuck on a stake into the bank it seemed misplaced because there was nothing there for an estate agent to sell, no houses, no shops or warehouses or offices, none of the usual things they advertise, only the fringe of trees, an undergrowth of brambles and fallen branches, the chain-link fence, the unused quarry. Perhaps what it was really offering can't be described easily on a board; a hole, 150 metres deep, said to be the largest man-made crater in Europe. What might anyone want with a silent, water-filled abyss surrounded by wire and warnings of danger? More than part of the city unseen, it seemed like the city's secret, a strange irony considering that it's where most of the city came from.

I began to think about what it would be like to buy the quarry. I reflected on the facts: *the largest man-made hole in Europe* (I tried to think of competitors but couldn't – the second-largest man-made hole in Europe?). Then I began to enumerate the difficulties. The first (after not having the money to buy it) would have been the Scottish system of purchase, which involves competitive bids in sealed envelopes and the common experience of discovering you've paid vastly more for the flat, the house, the quarry than you needed to. (Not only a quarry but an overinflated, expensive, disused quarry!) More important though would have been the unassailable fact that if I had bought it, I and only I would have been responsible for a vast, water-filled hole in the surface of the earth 150 metres deep, with a circumference of 120 metres.

The quarry is Rubislaw Quarry, where Professor Trail brought his students to collect plants when there were still fields and farms surrounding it, where truanting children went, dangerously, to play; a granite quarry worked for 200 years, with 30,000 tons of granite extracted, hewn from the seams of stone, hauled up from ever-lowering levels by teams of horses, the stone sent across the world to build bridges, ports, opera houses, courts, banks and used to build much of this grey Victorian city.

Granite is stone's stone. Its every construct and construction is adamantine. It is obdurate, its habit unyielding. Granite is not the stone you pick from a beach, a beguiling, rounded pebble. It is not the jagged splinter of white and glistening quartz that takes your eye, veined and moss-encrusted, the shard of amethyst you pick up on a hill. It's not the stone for reminiscence, to remind you in its touch for ever and silently of place, not the smooth stone that warms in your hand, the one that reveals its colour in a slew of rain. You can carve, but not easily, from granite. It doesn't generously provide soft and yielding limbs, soaring angels, gentle, weeping pietàs. Granite's for Amenhotep, for Ramses, for obelisks and monuments. It gives a cool grey dignity to its working. Marble seems to have light within, while granite protects its heart, sending only tiny mica sparks to glance capriciously from its surface. Granite is what my house is built from. I hold my hand against its walls, thinking of what and where this stone has been, 500 million years to come to this, through the Precambrian, the Cambrian, through the movement of tectonic plates, when Scotland was somewhere south, part of the continent of Laurentia, how this stone underneath my hand has been through volcanoes and glaciers, melted, fused, folded; of its components, quartz, mica, feldspar. (Radioactive too, it gives off radon, a potentially dangerous gas. The antidote in most granite-built houses is to open the windows, or to install a radon-sucking pump.) The walls of this house came from that crater, a quarter of a mile away. Different from sandstone, which softens, alters shade, mellows as it's washed and eaten by the substances in rain, granite is impervious to water and wind. Nothing weathers this stone. Granite darkens slightly from pollution in the air but doesn't wear or age.

Old photos show plunging cavernous depths that look natural, like a deep, steep canyon, a place created by time and not by man. The granite industry was harsh and hellish. Quarrymen worked in near darkness, in cold and damp once the workings were sufficiently deep. The quarry's sides dripped perpetually. Working with earth's most unyielding substance was injurious to life and health, a slow, or even a fast death by dust and grinding. After the installation of cable cranes called blondins, which lifted huge blocks of stone on thick steel cables, horses were no longer used for hauling.

The quarry was so deep that during the Second World War, the Department of Physics at Aberdeen University stored its stocks of radioactive material in a hut at the foot of the quarry to protect it from any bombing.

By 1970, the best seams were exhausted and the quarry was closed. For years, the slowly seeping water had been pumped away but now it wasn't and the deep hole began to fill. The chain-link fence was put up, the warning notices, the messages of danger. People forgot. Granite was imported from China. Oil company offices and flats were built above it on its periphery and broom grew on the bank above the road.

For days after I saw the estate agent's sign, I thought about it. I knew that for the person subject to systemic, persistent and habitual worrying, a water-filled quarry might be an unwise purchase. As it is, I dream of floods. I look out of the window in the rain at the ponds gathering at the corner of the street. I pass the Denburn where it flows under the road and listen to the volume of its rushing, and lean over the railings where I can try to gauge how much rain would be needed to fill it up to overflowing. I do this without owning the quarry. If I bought the quarry, I would dream of accidental drownings, of mass suicides, of horrors as yet unformed even in my febrile brain. I would imagine infestations, irruptions, algal blooms and toxic gases. I would have dreams of monsters rising, serpents, water-wraiths, the entire city inundated, water flowing through the streets, a quarry-led disaster movie. If I bought the quarry, I would never sleep again.

Knowing that didn't stop me from trying out a few small and fleeting dreams of totalitarianism. I would allow none of the potential developments cited by the newspapers as soon as the sale was made public. No climbing walls, no modern hotels, no more building. I would be ruler of that deep republic, mistress of the water and the air. I would rule my people-less country like an ancient, reading Mencius' enigmatic dialogues with the king, or, like Lear, shouting wisdom to the water and the air. Perhaps, in honour of my ownership, hawks would rise from their trees at my call and gulls would answer but I know they'd do that anyway.

As it was, within quite a short time the quarry was sold. My gratitude was boundless. I felt as if I had passed the burden of worry to someone else.

And now, on this July morning, we are gathered on the pavement by the wall. This time, there's no levering and scuttling. The gate in the chain-link fence is unlocked. A ladder has been brought and graciously, one by one, we make this short transition from the world of four-lane roads to one of near silence. We're as unlikely an assemblage as ever stood in the July rain on the edge of an unused, water-filled quarry anywhere. We're here to experience and to observe, an artist, a sculptor, a musician, a New York gallery owner, a photographer from Providence, the archivist for an Aberdeen museum and the owners of the quarry, Hugh Black and Sandy Whyte, and me. Some of us are here because we're part of an artists' residency at the university, the same one that took us to listen to birdsong in Seton Park.

Picking our way along the steep bank, we reach a point where we can all just about find room to stand among the fallen tree trunks, the softness of leaf mould underfoot, beech mould and bramble, cotoneaster and fern. The water is near and high. There's a scent of fungus, the only sound that of rain dripping into water. The height of the water against the banks takes away the sense of the depth below the surface, the steepness of the sides of this vast space. (Only the top branches of submerged trees hint at the rising of the water.) On the far side, one of the blondin cables that has been left in place rises from the depths. A line of black-headed gulls, twenty or so, stand along its length, impassive in the rain. (Why are they called black-headed gulls when their heads are the softest, darkest brown?) A few herring gulls fly overhead. Of the kestrels who used to nest here, there's no sign.

We stand under the trees, beech and sycamore, lime and rowan. The dark water floats with white feathers and leaves, bubbles with the force of falling rain. Clouds of wild roses and ragwort, rosebay willowherb and purple buddleia tumble from the almost perpendicular banks below the office blocks, which encircle the place like watchtowers. On the far side, it's all steep rock sides, there's no dry ground at all; this is the only place where you can stand.

The owners, local businessmen, Aberdonians both, have bought the quarry for its history, for its place in the life of the city. They want to preserve it and keep it as a nature reserve, to use it, at most perhaps for diving, to keep it for the city. They've put up a long measure on a tree to chart the rising of the surface. They would, they say, like to lower the water level. (How much is there? Where would the water go?) They are charming and cheerful and optimistic and appear happily resistant to my neurotic worry-transference.

Under the water, they say, down on the dug-out floor of the quarry far beneath us, is the machinery, the excavators and cranes lying there still and who knows what else, what has been dropped, or fallen, or been pushed in over the years since quarrying ended.

Looking and photographing, we edge along the bank, noting what we need and want to know, and then scramble up the wet bank and back down the ladder to the pavement. I don't know if we're all experiencing the same, if we're all feeling that in this place, something speaks to you silently and warns you of the danger of forgetting, that granite is stone that seems to measure us, seems to find us slight, negligible, altogether too little when measured against the length of time.

The Bird with the Silver Eyes

Friday's a sullen, lowering grey all day until evening when the sun begins unexpectedly to shine, illuminating both the road where Dave is cycling and a small bird hopping, frantic amid the traffic of a busy commuter route. When he phones me from the side of the road, it is with the greatest eagerness and delight: 'I've found your jackdaw,' he says.

I've wanted a jackdaw for years. I am acquainted with the commonest corvids but this one has eluded me. Every spring, I watch jackdaws playing on the roof opposite. I watch their pre-mating flight, their lascivious, tempting swoops and dives: I judge from their urgent busyness when nesting has begun, when eggs are being brooded but since this takes place in tall trees in the garden of an office across on the other side of a busy road, there is no chance of approaching closer. Jackdaws, for reasons I don't understand, appear to prefer not to walk about the streets of Aberdeen, or not the ones near me. I see their darting flights every day and hear their voices, but I've never met one face to beak. When I'm in Edinburgh, I encounter plenty from the colony that lives in the trees at the top of Bruntsfield Links, a large area of grassland in the middle of the city. Jackdaws strut and peck and fly and watch warily as they make their way among the students on bicycles and incipient jackdaw-stalkers like me but no fallen jackdaw infant has ever come my way, until this one. But I know at once this isn't a case of a bird who is learning to fly being mistaken for abandoned. It's far too late in the season for that.

Bec and Leah are visiting so Leah and I drive the mile to collect the bird. From under a fuzzy grey Mekon-mound of head feathers, a small face with a neat, short beak and watching, pale grey eyes peers out from between Dave's hands as he stands on the pavement waiting for us. The bird's tiny. It's unusual to see a jackdaw as small as this one in July. The nesting season's finished. There's only one fairly short period in a year when fledglings are about. Once it's over, it's over. But now, this out-of-time bird who, when I pick him up to put him into the box Leah holds on her knee, feels too delicate in my hand, his bones too near the surface of his skin even for an infant. The sharpness of his sternum, the powdery dullness of his feathers speak to me of disease or starvation or both; his size – that of a fledgling – seems wrong for the time of year, when others of this spring's *Corvus monedula* young are busily, darkly, glossily growing, chasing their parents with eager, flapping wings, still demanding to be fed although large enough to feed themselves.

The questions I asked Dave when he phoned about the whereabouts of the bird's parents and the proximity of trees seem irrelevant now. He's not a healthy, vigorous bird who has jostled himself from the safety of his nest, or one who has simply taken a fateful wrong turning while learning to fly. He is hatched too late, a starveling, ill.

My jackdaw. In spite of all my intentions to stop acquiring new birds, I hoped again this spring as I do every spring that someone would come across an abandoned fledgling jackdaw but as the season continued, my hope waned and the thought of a jackdaw was put into abeyance for another year. Jackdaws are careful nest-builders. They build lined, cup-shape structures, often in deep, well-protected spaces such as chimneys or dense trees, and I wonder if this is why there seem to be few young fallers-out-of-nests.

Leah, the jackdaw and I drive home. I transfer him to a large box, one I've kept since early spring in the spirit of expectation, or perhaps of hope. By late June I thought about folding it up and adding it to the recycling bag but didn't. A box is always useful. Now, I line it with newspaper and slot a small branch through the cardboard. The bird knows how to perch. He's almost fully feathered, and knows too how to fly. He does, this tiny creature who reminds me in flight more of a bee than a bird. He flies across the room to land on the sofa from where I scoop him up to put him back in his box. He pecks at my hands once or twice but soon gives up, whether from weariness or unearned trust, I don't know. I settle him in his box and take out the packets of food I keep in preparation for the spring arrival of a fledgling – the specially formulated mixes of everything a young corvid might want or need. As I hold the dropper with the food, I hope that the yeasty, malty scent of berries and whatever else there is in it might tempt him but he shows no interest. There's none of the frantic, eager hunger Chicken and Spike the magpie both showed as fledglings.

The bird is intriguing, different in many ways from the birds I know

247

best – rook, crow and magpie – in the set of his head, his stance and gait, the shortness of his beak, the hint of the panels of feathers which grace the faces of adults and most of all, his round and beautiful silver eyes. But he's still demonstrably one of the family Corvidae, utterly familiar in his look, his feet, his beak and feathers but more than that, in his smell, poignant and inimitable, the smell of a young corvid, one which Bec and I both recognise instantly and exclaim over. It's the smell we know from Chicken's babyhood, now more than twenty years ago, from Spike's too and more recently Ziki's, the kind of transporting smell which, as parents, we forget until in the presence of a baby we remember, when lost sensations are momentarily restored, the span of our own children's babyhood. (Han recently had occasion to deal with a near-drowned young swallow, rescued from an Italian lake. One of her swimming companions, a zoologist, put the bird in a paper bag until it recovered. On opening the bag to check on it, the first thing Han noticed, she said, was the overwhelmingly familiar scent of infant bird.)

In the jackdaw, we see the same fierce and all-encompassing gaze as in the rest of the corvids, the same watchful air of cool assessment, the same clear reflection of the workings of the same large, well-equipped corvid brain.

'This one's a genius,' Dave says, watching him. It's what we always say about them on our first meeting.

The jackdaw excretes, but won't eat. We dribble tiny quantities of liquidised food in through the sides of his beak, trying the manoeuvres and postures of parent birds to initiate the impulse that makes a young

248

bird want to feed. (Offering food to a now elderly Chicken in a particular way will still turn her from a very adult bird into a supplicating infant.) Yet anything he ingests, he does only because I manage by stealth to introduce some food into a briefly opened beak.

The jackdaw is interested in everything except food. I sit beside his box and watch him as he looks around, turns his small head, directs his silver eyes around him as if he requires to know and to see everything. I lift him onto his branch and close the flaps on his box.

'Twenty-five years have passed', Konrad Lorenz wrote in *King Solomon's Ring*, 'since the first jackdaw flew round the gables of Altenburg and I lost my heart to the bird with the silvery eyes.'

No one else, I believe, has ever known as much about jackdaws as Konrad Lorenz, Nobel Prize-winning founder of ethology, the study of animal behaviour. In the sixty years since *King Solomon's Ring* was published, no work has come close to the comprehensive authority of Lorenz's study, mention of which appears in virtually every paper and thesis on any subject relating to jackdaws. He undoubtedly enjoyed certain advantages over most of the rest of the world of assorted bird-studiers and ethologists, living, as he did, in a mansion with large grounds near Altenburg in Austria, in the quiet, virtually traffic-free days between the two wars when it was possible to cycle around a small town with jackdaws in pursuit, or sitting on his shoulder. In the preface to the book, he writes lyrically about the beauty of the countryside surrounding Altenburg, the marshes and reed beds of the Danube. The other advantages he appears to have enjoyed were a sympathetic family and the space to study shrews, geese, dogs and everything else he gathered around him. (It's not everyone, I imagine, who had the resources to rear fourteen jackdaws in accommodation easily adapted for experiments in jackdaw social attachment.)

Briefly, I put aside my anxiety for the bird and begin to feel anxiety for myself, the kind you feel when you have just acquired a new and potentially long-term responsibility. Where will he live? The study and kitchen are Chicken's domain, the rat room Ziki's. Could the sitting room be his? He seems too active to contain in one – albeit large – room. In any case, I can't isolate him. Can I train him to come when called, to sit on my shoulder? What will Chicken think of him? (Not, if I know her, very much.) And Ziki? How soon can I introduce the jackdaw to them? How soon before I know if he's free from an infection that could be passed to them? I remember that I have to travel to Edinburgh in a few days' time. Can I take him with me? Will he be frightened in the car? How will I transport him? Is he too fragile to travel? Where will he sleep? The logistics of housing him might be difficult but not insurmountable. Within a very short time, friends begin to offer birdcages, an antique parrot-cage, a cage once used by a dog.

When talked to gently, the bird replies, an infant version of an adult jackdaw's speech.

The bird won't eat before bedtime. I carry him upstairs in his box and put him beside the bed. It's one of the few warm nights of this summer. I wake several times and check on him in the darkness, opening the corner of the box, looking at him in an edge of torchlight. Then I lie and listen to him stirring, moving on his branch and wonder how this bird from the most gregarious of societies feels, alone.

He won't necessarily be afraid. Young jackdaws are not predictable in their fears and have no inborn fear of predators. In this, they are unique among corvids. They don't fly away from the things of which other young birds are rightly scared – cats, foxes or squirrels. According to Lorenz, they have to be instructed by their elders 'by actual tradition, by the handing-down of personal experience from one generation to another', their only innate fear reaction being inspired by seeing any living creature carrying a black object, particularly if it's limp or flapping, when a furious jackdaw mob might surround the unfortunate, unwary carrier of a dustbin bag or black scarf. Like other corvids, jackdaws have long memories and a tendency towards enduring grudges. The person once seen carrying an object deemed offensive will never again be rehabilitated into the good offices of jackdaw opinion. (Crows, too, demonstrate the same ability to maintain a grievance. Biologists who have ringed crows for study describe finding themselves picked out of crowds of thousands on their university campuses, to be pursued by shouting crows.)

Jackdaws, or Eurasian jackdaws, *Corvus monedula*, called in the United States 'Western jackdaws', live in an ordered society of strict rank hierarchies, with dominant males acting aggressively only towards those of slightly lower status. Lorenz describes high-caste jackdaws as acting with condescension towards those of the lowest rank, regarding them as 'merely the dust beneath their feet' although Lorenz praises the ways in which the social order is respected and maintained. The act of pairing is one of the only ways to alter the status of a bird within the ranking. (Interestingly, only females can improve their status by these

251

means. Lower-order females may improve their status by mating with higher-order males. The reverse, apparently, is never the case.)

One of the most lovely, and haunting, descriptions of the relationship between a human and a bird is in Elspeth Barker's book *O Caledonia*, in which the splendid young heroine Janet establishes a bond of such closeness and love with her pet jackdaw, Claws, that after her death, in inconsolable grief, the bird flies against a wall and kills himself.

The mating and pair-bonding of jackdaws is singular. As most jackdaws do, this bird's parents would have – to use the words of Konrad Lorenz – become betrothed in their first year and married in their second. (Writing of the relationships between male and female jackdaws, Lorenz is conscious of the possible criticism of his use of the terms 'falling in love', 'betrothal' and 'marriage'. The first of the ethologists – those who study animal behaviour, allowing and investigating ideas often regarded as anthropomorphic – Lorenz was a pioneer, laying down the foundation for future researchers such as Jane Goodall.) Jackdaws are long-lived birds who form enduring, exclusive partnerships, and although studies have demonstrated that in some avian species the perception of monogamy is in fact incorrect, with jackdaws it appears to be true. Lorenz describes the mating behaviour of jackdaws, talking of 'love at first sight', of the exchange of languorous looks and whispers, of the jealousies, *tendresses*, the desires to impress – in short, all the delights and follies of love. Writing at a time before DNA testing, his thesis was based on observation alone but the truth of his work is borne out by research carried out in 2000. A DNA study of a colony of jackdaws demonstrated that, in spite of opportunities for extra-pair copulation, there was not a single example of extra-pair fertilisation within the colony. Lorenz witnessed the dissolution of only one jackdaw 'marriage', but during the betrothal and not after the marriage. When jackdaws tie the knot, it's tied.

In the morning, at six, I prepare more food, hoping that at last the bird will be hungry, that he'll open his very small beak and flap and beg but he's no more interested in food than he was the day before. While I try to feed him, he turns away from the dropper, shakes his head, pebble-dashing the wall, the rug and me with baby-bird mush before leaping from his box to fly up to the chest of drawers, where he lands beside the model crow made of shaved balsa wood given to me by friends. They're exactly the same size, a model crow and a young jackdaw, 14 centimetres from the top of their head to their small black feet. He stands for a moment, looking with brief unworried interest at the model bird before he flies across the room to land on the bookshelf. He's active, busy, interested in spite of his lack of appetite. Perhaps he's just adjusting. Perhaps he'll rally, become hungry, eat. I can't allow him to use his energy in flight if he won't and so, reluctantly, I put him back into his box.

Too-late hatched. Jackdaws' eggs hatch asynchronously – not all at the same time – the earlier-hatched young having a survival advantage over the later-hatched, who often are left to die. In jackdaws, 50 per cent 'brood reduction' is common too – the birds lay more eggs than they can rear and while the first to be hatched benefit from any available food, the later ones may either be killed by their siblings or starved. (Both are methods of ensuring brood survival when food sources are uncertain. These strategies are what make jackdaws 'successful', help to maintain their numbers, keep them on the 'least endangered' lists.)

The day is warm. We carry box and bird downstairs and install them in the sitting room, where the small bird won't overheat. It's a room with a high ceiling and always cool. (Granite holds the chill, seeming to emanate slow memories of winter from the walls even on the hottest days.) We spend as long with the bird as we can, talking to him, trying to encourage him to eat, sitting by his box. In response, he jumps and leaps and wants freedom. When, for a few moments, I let him fly freely, he takes off towards the bay window and the rocking horse that stands in front of it. Landing lightly on its saddle, he stands as a small emperor might, to inspect a crowd. He stares around him with the pouting, aggrieved look frequently assumed by those still in possession of the frilled gape of infancy. I leave him for a few minutes so that he can stand, turn, look out of the window at the leaves and sun.

Jackdaws' eyes are utterly different from the eyes of rooks, crows and magpies, whose pupils are various shades of dark brown and irises black. Uniquely, jackdaws have black pupils and pale, startling silver irises. In an article published in *Current Biology* in 2009, Nathan Emery and Auguste von Bayern from the Department of Animal Behaviour at Cambridge University reported on research demonstrating that jackdaws have a greater ability to interpret human visual clues, such as the direction and change of gaze, as well as other gestures of communication such as pointing, than most other species, including dogs and chimpanzees. It may be that the similarity of human and jackdaw eyes makes such identification of visual stimulus possible and that the demands of long-term pair-bonding encourage greater awareness, sensitivity and acuity in the interpretation of eye movements. The jackdaw's curiosity seems to draw everything into the scope of those intense and glittering eyes.

The evening is one of the few in a northern year when it's warm enough to sit outside in the garden. We come in every half-hour or so to visit the bird. Still, he won't eat.

While I'm sitting in the garden, I watch a female blackbird for a long time. Her preening is scrupulous, concentrated; her stopping on a branch is a pause in a life both busy and demanding. Her life, I know, is no less remarkable than any other creature's, bird or human.

In the evening's heat, the swifts scream over the garden. Towards darkness, we take the jackdaw from his box and hold him and stroke his feathers. He seems content. Bec puts him on her shoulder and as he sits there, his eyes grow heavy. We know that it isn't the night-time weariness of a busy baby bird. He folds his head under his wing and when we put him in his box, we know that it is the last time we will. I had hoped that his vibrant desire to know and to see might be powerful enough to overcome his physical weakness but I recognise that this thought is part of an unfathomable human optimism, a mistaken belief that the will can overcome all adversity, the processes of disease in both man or bird. I take him upstairs with me again and put his box by the bed.

I wake at four to check on him and, on lifting the flap, see him lying, a small black wing unfurled against the newspaper lining his box. I had hoped he'd survive, while knowing that he wouldn't.

This morning, I bury him between the glaucous-leaved hostas in the east-facing flowerbed and as I do, acknowledge that his will was very great.

MIDSUMMER

July 17th

There have been a few warm days. I kneel by the pond to watch whirligigs and diving beetles, pond skaters, water boatmen, the common froghoppers disgorging their transparent froth, 'cuckoo spit', on the stem of one of the tall grasses growing from the stones. Through the walls of ferns, the sound of splashing as a young male blackbird bathes.

Not only in the water but all around me birds are bathing in the sun. On the table, the ground, the chairs, blackbirds and sparrows are sprawled in the posture we call 'dying rook'. (The term was invented after our first observations of our young rook in sunshine. Stricken by panic at the strange sight, we quickly made the gloomiest assumption. It was only when she shook herself, resumed her normal posture and began accepting fat aphids we fed her from the roses that we realised she was perfectly OK.) On hot days, you'll see birds spread on bushes, on grass, anywhere where there's sun, one wing outstretched, head lolling, beak often alarmingly hanging open, eyes closed, looking for all the world as if death is imminent. It's not. They are absorbing what vitamin D they can. Feathers are difficult things to get the light past – they have to be super efficient to keep out the cold.

A collared dove too is performing 'dying rook' on the precarious curved black metal arm of the anti-squirrel bird-feeder, its wing feathers lined up, neatly spread out like a perfectly tailored set of pleats, a bird with wings by a Japanese designer.

The leaves of the small acer by my side shiver, burn red in the sunlight. Behind me, in the thicket of branches, the sounds of voices and wings. Swifts shriek overhead, and beyond the walls, I hear the sounds of traffic.

We dug the pond after we'd been here for a while, lined it, surrounded it with stones and pebbles and put in tadpoles. At first, when it was new and the water clear and unclouded, I used to watch the growing, thriving frogs through binoculars from the study window during interludes in my working day. I'd learned that, even if I crept down the lawn, they'd see my approach in shadow and disappear swiftly into the mud and stones. Through the lenses I'd see their eager heads and watching eyes. I planted ferns and iris and water lilies that floated delicately on the surface of the water. The doves bathed in it in spite of having their own bathtub outside their house. They pottered on the grass and among the stones of the paths and in spring, nibbled the tops of young bulbs before they appeared.

After a few years I decided to clean the pond, filling buckets with its water, intent on scraping up the silt of old leaves and rotted crab apples, the viscous mud and tangle of dead roots. As I was scooping up the last of the water, I saw the face of a small frog watching me from the puddle that was left, a creature in an element destroyed. I hadn't known that there were any frogs left. Quickly, I poured the water from the assembled buckets back again and hoped that frogs have short memories and that they don't remember random, purposeless assaults upon their personal space.

When, some time later, I found a frog one evening on the back step, I wasn't sure if it was the same one. Frogs are difficult to tell apart. I carried him down the garden and slid him onto a rounded stone from where he leaped into the darkness of the pond.

The pond frogs were small, few and quiet, *Rana temporaria*, common garden frogs. I wished they were there in loud and croaking numbers, shouty, vehement, hot-weather frogs of the kind who once, years ago, sang in marvellous concert with the world-renowned violinist who had come to entertain the members of the kibbutz where I was living. He played while the frogs sang accompaniment from their ornamental pond outside the recital room, overwhelming almost to inaudibility the beautiful, careful rendition of a Bach partita being performed inside. (They would have been marsh frogs, *Pelophylax ridibundus*, or green toads, *Bufo viridis*. There would have been more then than there are today – I read that amphibian numbers have fallen in Israel as they have everywhere, as they have here, for the usual reasons: pesticides and building, loss of habitat, disease.) The audience and the famous violinist silently trembled with suppressed, delighted laughter.

Once, a young heron drifted over the gardens behind ours and across our wall to touch down lightly beside the pond. It stood only for a moment before it took off and disappeared again like a long, grey shadow.

Over time, the pond has leaked water and become more of a small bog although there's still sufficient water for things to live and bathe. The lining, I suspect, has disintegrated in parts or has been unable to resist the strength of the roots of the water plants. Behind it, the east wall has been left to itself so that the espaliered apple and pear have grown far beyond their margins.

On the leaves of the cotoneaster I find a small coral-pink beetle and capture it briefly. Its wings are pearled, its eyes black pinpoints. It is possibly Hemiptera, *Psallus varians*? (Is it important for me to identify this tiny creature correctly? A guide to Hemiptera tells me that attempts to identify this species by means of external features are often impossible and so even with a microscope, I couldn't be sure.) Whatever he is, he's beautiful, running over the centimetre scale on the bottom of my viewing jar. I watch him for a few moments and wonder what his time frame is in relation my own, and then I release him where I found him.

Reading the other day about a scheme for recording sightings of mayflies in Scotland, I discover that they're of the order Ephemeroptera: winged, short-lived, living for a day. It seems to describe all life; the perfect kind of metaphor.

The first chanterelles have appeared, or the first that Dave has picked this year. Over the years he's been collecting, he's noticed that the mushroom season has advanced by weeks over the past two decades. He has collecting areas in various woodlands all around the city although I don't know where because mushroom-collecting is a strangely secretive, defensive art. With a damp summer, there'll be huge bags of the frilled orange fungi in the fridge for the next few months.

On these evenings, the not-quite-white nights of summer when dusk is late, bats fly over the roof, round the corner, back again, too fast for me to catch in the lenses of my binoculars. They're pipistrelles but I'd need a bat detector to know which species they are, common or soprano. The difference is simple to explain: the soprano pipistrelle's echolocation call is higher than that of the common pipistrelle. The summer's rain will be disastrous for baby bats who are just learning to fly.

It's difficult to imagine but these lovely, fleeting, fascinating beats of darting shade are remarkably long-lived – they can live for 40 years, much longer than most other mammals of their size. There are brown long-eared bats in the area too, and Daubenton's bats but fewer than when William MacGillivray wrote his paper 'Description of *Vespertilio daubentonii* from Specimens found in Aberdeenshire' in 1844. The bat box we put up years ago has fallen from its mooring on the beech tree and I haven't fixed it back up again.

July 18th

Heavy rain is forecast again on a day as dark as November. Late on it falls, filling the gutters, then turns to a spray so fine that when I see it blowing past the streetlight, I think it's smoke. I think of the saturated doo'cot floor, and of the baking streets of Europe's cities. I wait for a few hot days when I can scrub every surface of the doo'cot as I used to do a couple of times each summer, with the certainty that everything would dry before dusk, but the days don't happen.

July 20th

This morning, after checking Aurora Watch, I read a message on NASA's website suggesting that astronauts should watch out for auroras. As the 30-year-old Atlantis space shuttle programme is coming to an end, the powerful interaction of solar wind and earth's magnetic field is likely to lead to a 30 per cent chance of high-latitude geomagnetic activity. It suggests that looking out of the window of Atlantis could be a rewarding experience. And they've only got a day to see it. I think of them watching, floating, circling then diving towards earth. Here on earth, at this high latitude, it's raining. The shuttle returns but here, there's still no aurora.

I think of the astronauts and remember a night many years ago, before we were aware that we had to calculate the effects of our flying and weigh need against harm, when I flew south and east on a cloudless November night. Half the world's sky was cloudless, vitreous and starlit. I've never forgotten it, seeing the ordering of a segment of the planet in detail below. We flew over Europe, all neatly structured, cities flickering sequentially like wide nets of airy brightness, then over sea and sudden darkness. Desert was blackness bisected by two parallel road lines of scorching brilliance, the only light as night stretched across continents. It was a map alive. The proximities of lights explained how people lived and how far they had to travel. It looked like an embroidered landscape, Iranian towns like tiny spills of sequins in the darkness, the villages of Rajasthan where lights were tight clusters of French knots sewn over blackness. I didn't sleep for watching. This wasn't land interrupted, the interminable six hours of sea then land again, of flying to America. I couldn't close my eyes. Our path was tracked across the small screen in front of me, over countries and history and time. First light was a luminous, watered dawn over the Bay of Bengal; pale blue haze over sea and spidery inlets, low islands in a vast wash of broken water. I measure my sense of the earth from that night world as it appeared to me during those hours, a synergistic view of people and place, understanding in a way I hadn't done before the delicacy and wonder of the planet. Describing the first pictures of earth taken from the moon, humankind's first sight of the entirety of its planet, American ecologist the late Eugene Odum described it as looking 'so fragile, so small and so alone'.

July 23rd

This morning, with the usual expectation of disappointment, I check Aurora Watch and lo, there in a box of yellow appear the words 'minor geomagnetic activity'. I can hardly believe this manifestation, if not from heaven then from the magnetosphere. I have faith in this most welcome prediction, as I have faith in science. I consider breaking into a chorus of the 'Northern Lights of Old Aberdeen' but don't, just in case. I check the weather, the other set of pictograms, those little boxes in which, on the BBC weather website, the information is brought to us, the symbols of cheer or gloom indicating that the sky on this particular evening will be clear. I prepare to alert all the known aurora hopefuls of my acquaintance and all my Aberdonian friends in other places, even the ones who will have no hope of seeing it but would be rightly jealous, or nostalgic.

Later in the day, I check again. Whatever sign manifested itself to SAMNET, manifests no more. The band has returned to green and carries the usual message 'No significant activity'. The minor geomagnetic activity has gone away, melted back perhaps into the fathomless skies. In case a mistake has been made, an over-hasty dismissal of the possibility, all evening I check the sky. It gets dark late. By ten, the sky is still light and cloudless. Official time of sunset is just before eleven and the sky still glows, backlit by brightness in the northern sky. I go into the garden. It's not quite dark but I look up. There's no sign of the aurora but I wait on for darkness – it's a Saturday evening, and I think of the ritual at the ending of the Shabbat that happens with the first sighting of three stars in the sky. It's called *havdalah*, from the Hebrew 'to separate' because it separates Shabbat from the rest of the week but I've seen it carried out only once or twice. Strange and beautiful, it's meant to involve all the senses, with wine and candles and the scent of spices. A part of it is that you have to hold your fingers out towards the

266

candlelight, to see the glow of light reflected from your fingernails to remind you to appreciate the smallest aspects of your life. I breathe in the scents of damp earth and leaves. Watching the sky, I count the first three stars and as darkness approaches, although there's no sign of an aurora, I try to identify the great constellations visible in a northern summer sky.

July 26th

This evening, driving home from visiting friends, I stop by the Ythan Estuary, a few miles from the city. It's one of the most important European sites for wading birds and migrators. An evening after rain, the light is hazy after a mild, damp day. Even the sharp, acidic yellow of growing rapeseed is muted, its brash squares mellow among the pale blues and greens of sky and fields and trees. It's chilly as I leave the car. Through binoculars, I see four large white birds feeding at the estuary, tall birds with spatulate beaks. I don't know what they are but some other people who are watching too, identify them. They're spoonbills, very rarely seen here. There are herons too, a greenshank, oyster-catchers. The spoonbill, *Platalea leucorodia,* is an almost silent bird. It's such an odd idea. 'Grunts during the breeding season' is the enigmatic comment on the bird's limited vocal repertoire in one bird book. Driving home, my cold fingers thaw slowly. There have been only a few days of true warmth all summer.

Flying through the Storm

The garden flutters, alive with threatened birds. On a lowering morning in August, a morning almost magnificent in its dark-skied, raining dismalness, they're gathered here outside my window, unceasing in their busy fortitude, apparently impervious to weather, feeding at the bird-feeder, crowding in, jostling, nudging, singing; beautiful, small, vital, ever-moving, their voices sparking brilliance into the blue-grey darkness of the morning. *Passer domesticus,* the house sparrow, like busy, mobile fairy lights garlanding the overgrown viburnum and philadelphus. Even unseen, they're there in the sway and dip of branches, in a sudden chasing, a sudden domestic (or *domesticus*) quarrel or avian outrage of one sort or another, in an outbreak of loud and shrill complaint. An incident of passion or violence becomes an explosion of leaves, a bout of vigorous grooming and preening transforms into a scatter of raindrops from the branches.

In spring, the garden is filled with nesting sparrows, with rushings to and from the depths of the ivy, the carrying of building materials and then food and then, after a time, fledglings. Soon, they're all over the bird-feeders, calling; they're on the backs of the garden chairs, on the surface of the table, quivering, fluttering, begging. They appear suddenly, entire families of them, like a Busby Berkeley chorus in unison from the newly growing spring arrases of green, to feed and hop and sing and flap and quiver before retiring back as one into the thicket of rose, jasmine and pear. On bright days, they sit, peering, as from windows on their individual branches, chattering, calling, like a noisy shrubful of watching concierges.

Their singing is loud but if something stops it, it stops suddenly, as if it's been switched off. Then we wait, indoors and out, in silence until whatever danger they have discerned but I haven't has gone and those of us who sing, resume our singing. Sometimes, I walk from the back door to fill the bird-feeders into the shrill and shriek of sparrow alarm, a sound that sparks like fire through the garden, a shrub-to-shrub, tree-to-tree incendiary, small, high choirs of warning. Usually, I'm just too late to see them fly and so I know that I'm not causing their alarm. I rely on their judgement and their observation because they're never wrong. When I look for a cause, I find it – the neighbour's cat watching us all from a window, a burst of alarm-calling from another garden as a hawk flies overhead.

Sparrows are a constant presence in the garden now as they weren't a few years ago. I don't know if it is because the garden has grown so abundantly that it provides more places for them to nest and to feed, or if my casual gardening methods have encouraged insects, or if it might be as a result of a northward move of populations.

I appreciate their being here and the knowledge that these birds, in their almost inexplicably diminishing numbers, with a vast range of possibilities for other places to live and feed, have chosen my garden and my bird-feeders.

Before the snow began I bought a 'sparrow terrace', a wooden structure with sparrow-sized doors to welcome them and encourage them to breed. I had to wait for weeks until the snow had melted when it could be hung on the outside wall, high, above the ivy where the doves once liked to roost. Sparrows, apparently, like facing east (is there a sparrow Mecca? A *domesticus* Jerusalem?), so now it faces east towards the sea. I have to crane to watch them, separated from them only by the glass of the window, the now endangered and the future endangered; for none of us may be as certain as once we were of our future on this earth and as I watch, I know that I'm a representative of the species which has caused their endangerment. I know that their presence indicates nothing about the wider situation of decline but in Britain the house sparrow is a threatened species (as are the herring gull and the starling, too). They are 'red list' birds, ones we forgot to value or didn't know how to, ones we didn't know we needed to protect.

Sparrows live near us and with us, as the American writer Sandra Steingraber writes: '. . . they cannot live without us. They are our avian shadows. They are Ruth to our Naomi. *Wherever you go, there shall I follow. Your home shall be mine.*'

Henry David Thoreau valued them sufficiently to write that having a sparrow alight on his shoulder while he was working in a village garden distinguished him more than being awarded any epaulette.

'Proletarian birds,' the poet Norman MacCaig calls sparrows

his taste in clothes is more
dowdy than gaudy
And his nest – that blackbird, writing
pretty scrolls on the air with the gold nib of his beak
Would call it a slum.

I don't think house sparrows' clothes are dowdy; these small, brown, not drab, not dowdy, not dull birds, these birds with plumage the colours of earth and autumn. How could something as quick and bright be dowdy? House sparrows, known by those who think that common, or even once common, creatures should be called by schoolboyish, scatological names to show how much beneath notice and interest they are, as *Little brown jobs*. The birds in my garden are all common, unless I've failed to notice the occasional rarity. I like common, ordinary birds, the latter with its implication of routine, the pleasant quotidian calm of it. (To hear birds dismissed or described as ordinary makes me feel the same as when I hear human beings describe others, loftily, as 'ordinary' as if, by some process of mystical elevation, which should be obvious to us all, they themselves are not.) There are other species of small brown birds in the garden, blue and yellow birds too, black birds and black and white birds and dotted-all-over birds, birds with some red feathers and some green, but not bright red ones, nor luminescent green (unless the odd starling touches down, but a starling's feathers are only gorgeous up close). All are urban birds of the sort most often seen in any northern British garden although some of them are, by virtue of something we don't yet know, lessening drastically in number.

271

Sparrows seem a paradigm for humanity's contrary view of creatures: noticed and valued only when, for reasons unknown or only partially explained, they have embarked on the process of lessening their long hold on this earth. Birds that once were so numerous, so common (and so injurious to crops and grain stores) that they were trapped and killed in unimaginable numbers often by 'sparrow clubs' organised in British cities to destroy them (sometimes expanded in their remit to 'rat and sparrow' clubs). Roger Lovegrove, in his book *Silent Fields: The Long Decline of a Nation's Wildlife*, estimates that in England, 100 million sparrows were killed between 1700 and 1930.

In spite of their massed presence in this garden, sparrow numbers have dropped in Britain by 71 per cent over 30 years and no one really knows why. In Edinburgh, sparrow numbers are down by 90 per cent. Numbers have fallen in other places, too – in Prague by 60 per cent, in Paris by 10 per cent. Belgrade's sparrows are disappearing. A friend who lives in Amsterdam tells me that ten years ago, all their sparrows vanished mysteriously. (Renewed roof tiles were implicated.) She tells me though that this summer, Berlin's and Dresden's sparrows seem to be thriving, apparently abundant everywhere, singing above the traffic and eating from her hand. A splendid transcontinental sparrow-observer, she reports that in Lundy Island too they appear well and thriving. In India, sparrow numbers have declined so much that a World House Sparrow Day is held to encourage people to provide food and water and nesting places in their gardens and on their balconies.

Theories about the decline of sparrows are abundant: sparrowhawk predation, the intervention of magpies or cats, the effects of mobile-phone masts or other varied forms of pollution, the intensification of agriculture, the loss of nest sites often due to new building methods, the paving over of gardens, disease, the increase in the planting of Leylandii hedges (and other plants birds don't like), or all of them combined. Many people have worked hard to find out why, especially after the *Independent* newspaper offered a reward in 2000 to the person who could provide a convincing explanation for the phenomenon. The explanation had to be one published in a peer-reviewed scientific journal, and the competition was to be judged by representatives from the leading ornithological organisations and by Dr J. Denis Summers-Smith, a peerless sparrow expert and author of, among others books, the authoritative Poyser monograph, *In Search of Sparrows*.

In 2008, Dr Kate Vincent of De Montfort University submitted a paper suggesting that the decline was as a result of sparrow chicks starving from lack of insects. The still unclaimed prize was nearly awarded to her, but the jury was split in its judgement. (Dr Summers-Smith believes the decline to be a result of atmospheric pollution.)

Mobile-phone masts seem to be among the least likely reasons because the decline in sparrow numbers began before mobile phones were common, and hasn't been as marked in some places where masts are abundant. Cats, the major predator of birds and small creatures in British cities, may play some part in the decline but the drop in numbers is so great that it is unlikely they are responsible, particularly because there will be cats in areas where sparrow numbers have not declined. Magpies too have been discounted as serious contenders –

they've been shown, despite all belief to the contrary, to have no deleterious effect on numbers of other avian species. Possibly the most
acrimonious of the debates on the subject is between the opponents
and defenders of *Accipiter nisus*, the Eurasian sparrowhawk, a name
derived from the Anglo-Saxon *spearwa*, sparrow, and *hafoc*, hawk. In
the years after the Second World War, the use of agricultural
organophosphate pesticides led to a significant decline in sparrowhawk
populations, reversed after the ban on DDT and aldrin, a reversal that
may or may not correlate with the decline in sparrow numbers.

My own acquaintance with these splendid raptors has been almost
as long as my acquaintance with doves (the timing, I believe, is not
coincidental) but I had never seen a hawk either chase or catch a sparrow until a morning in spring while I was sitting at my desk, when my
attention was diverted suddenly towards movement outside, something
approaching fast, a bird whose progress raised the collared doves briskly
from their branches by its low and stealthy flight. Outside the study
window, a sparrowhawk landed to perch on the bird-feeder a few feet
away from me. Leaning forwards for a moment, he looked around with
his glorious, scary golden eyes. The sparrows who had been thronging
the feeder percipiently enough melted speedily into air or the branches
of the viburnum. Sparrow-less quiet suddenly surrounded this splendidly striped and lovely bird. What was he doing chasing sparrows? I'm
used to seeing sparrowhawks dining more heartily, ripping into plump,
succulent doves, ones I know to have been fed on the best grain assortment available to man or bird. I wondered if, like a diner in a restaurant
ordering a plate of insubstantial amuse-bouches, a sparrow might be a

small preparation for the more substantial fare to come. Surely it could hardly be worth the effort of plucking? (Sparrowhawks are neat and thorough pluckers.) But in the case of this hawk, the amuse-bouche had flown. As I stood up cautiously to look at him more closely, he took off from his perch and glided slowly away, low under the arch of ivy. I was sorry that he'd had to leave like that, empty-beaked. He was so beautiful, his flight so graceful and measured, a lesson in purpose, niche and adaptation. Within moments of his going, the sparrows were back.

Commonplace these small birds may be, but wait long enough and the ordinary becomes extraordinary. Familiarity, geography and scarcity all affect our attitudes towards other species, something I realised when my Australian mother-in-law first met our cockatiel and our rosella, both very common Australian birds. We thought them exotically beautiful and delighted in their appearance and their song, in their cleverness, their complexity, the relationships we had with them while, thinking only of their ubiquity and their noise, she was heard frequently, *even in their presence*, to mutter the word 'pests' under her breath.

Daily, as I watch the sparrows outside the window, I think about what the decline in their numbers might mean. Would it matter if these birds disappeared? Evolution deals with many species, extinctions occur and have occurred during the almost unfathomable lengths and breadths of time, but this isn't evolution: it's the effects of the actions of man destroying the futures of these birds and countless other species. The Species Survival Commission for the IUCN – the International Union for the Conservation of Nature – suggests that humans are causing species extinctions faster than new ones can evolve. On their list of 'threatened' taxa are 21 per cent of named species of mammals, 13 per cent of birds, 30 per cent of amphibians and 28 per cent of corals. The American environmental writer Bill McKibben writes chillingly in his book *Eaarth*, 'We're running Genesis backward, de-creating.'

If we ever imagined that numbers would ensure the future of a species, we need to remember the passenger pigeon, possibly the most famous extinction resulting from the actions of man. Almost unimaginably abundant, numbered in billions and flying in vast flocks which, according to John James Audubon, took three days to pass overhead, during the latter half of the nineteenth century passenger pigeons were netted, trapped, shot and hunted in an orgy of ferocious human cruelty and, as a species, utterly destroyed, removed entirely from the earth. In an article first published in 1924 and reprinted in the anthology *At Home on This Earth*, the writer Gene Stratton Porter writes in 'The Last Passenger Pigeon' of her childhood in the late nineteenth century when nature seemed so abundant that man 'waged destruction without a thought', as he did against this unfortunate bird. She describes her childhood love of the bird, its call that sounded like the words: 'See! See!' and of her experience of visiting a neighbour after a hunt when passenger pigeons were being prepared for the pot:

> I was shocked and horrified to see dozens of these beautiful birds, perhaps half of them still alive, struggling about with broken wings, backs, and legs, waiting to be skinned, split down the back and dropped into the pot-pie-kettle.

On telling her father, who wouldn't allow the killing of any pigeon or dove, believing them to have a special religious significance, he foresaw that – unlikely though it seemed at the time – if the birds continued to be killed in such numbers, they would be destroyed. He was correct. By 1900, there were none left in the wild. In 1912, Porter, alerted by someone to a nest of an unfamiliar bird in the territory of Indiana where she photographed birds, she found what she confidently recognised as a passenger pigeon, a sole male. Familiar as she was with the bird from her childhood, she recognised first its call and then its appearance. (This account appears to be the final recorded sighting of the bird.) She knew that what she had found was the last of the last. She listened to its call in which she heard a cry of bitter admonition, 'See! See! See!'

Standing with an air of slight self-consciousness in a glass case in the Zoology Museum at Aberdeen University, there is a passenger pigeon. A pretty, dove-ish bird, slightly frowsy as one might well be with stuffing and time, it is an enduring reminder of the folly and vicious stupidity of which man is capable.

In this museum, you feel close not only to the history of the specimen but to the species itself, so near and open to inspection. Birds stand in their modest, naked skeletons or in feathers displaying the wear of age. Creatures of all sorts collected long ago pose gently against the painted backdrops of their cases next to serendipitous collections of fish jaws and skulls. Here, you feel close to Linnaeus and to Darwin and to all the others who so assiduously travelled, collected, described: to William MacGillivray, who was born only metres away; to his friend John James Audubon whose gift of the skins of twelve North American birds is in the university's collection. Some of them lie in a glass case, saved by the sharp-eyed person who, in 1954, hesitated before throwing them out as yet another heap of dusty rubbish and on reading their labels saw, written in William MacGillivray's own hand, 'A present from Mr Audubon'. There is an Eastern meadowlark, a dark-eyed junco, a savannah sparrow, a yellow-throated vireo and a pied flycatcher, labelled round their legs, their curled feet looking like tiny wringing hands. (Audubon, it must be said, was a man who appeared to have a fondness, if not a positive enthusiasm for killing, not only the birds for whom he professed a sense of 'intimacy' and whom he portrayed in such number and detail, but a variety of other creatures as well. The American poet Mary Oliver writes of the dangers of the too earnest pursuit of knowledge, of Thoreau gassing moths, of Audubon thrusting a needle into a bittern's heart.) Among the bird skeletons is a sparrowhawk, oddly small when stripped of his feathers, flesh and life, this terror-inducing bird diminished in his power, standing meekly in his bones.

Sparrows are important in unimagined ways. In 1958, that master of both intended and unintended consequences, Mao Tse-tung, instituted, along with other measures during the economically and socially catastrophic Great Leap Forward, a campaign to kill pests. The pests were mosquitoes, flies, rats and sparrows. All were pursued vigorously and destroyed. My Chinese teacher at the Hebrew University, raised in Beijing during the most turbulent and terrifying years of China's modern history, described the results of the killing of sparrows, the clouds of flies which – being more difficult to exterminate – proliferated and swarmed, unhindered by the sparrows who had kept their numbers down, a torment added to many others being endured by that unfortunate nation. She told of the terrifying thinness of everyone she saw, drawing her fingers down her face to suggest the gauntness of starvation, a result of the drastic famine estimated to have caused 36 million deaths between 1958 and 1961, caused at least in part by the loss of crops to locusts which multiplied in the absence of sparrows. (The human story was as fascinating as the avian, or political – my teacher, Dora Shickman, a Han Chinese, a woman of great beauty and erudition, had managed with fortitude and suffering to escape from China during the Cultural Revolution with her Russian Jewish husband, to settle eventually in Israel.)

On a smaller scale, it is easy to effect apparently minor but noticeable change. So often, as with the removal of Aberdeen's starlings, we act without judging our future loss. Some years ago a nearby street tree was cut down. A small, undistinguished tree it may have been, but it was the tree where in autumn, fieldfares chose to gather in astonishing numbers, where they crowded, paused and drew breath before flying to consume the berries on the cotoneasters and rowans in nearby gardens. Since the tree has gone, there have been fewer fieldfares. I see them but they're scattered among trees down the lane, dispersed, not in the calling, exuberant, glorious chattering cloud there was.

I see changes in the district. I notice them as I walk or run; another garden gone, another garage built, another monobloc driveway destroying the places where infinite numbers of creatures once lived in the earth of the lawn; in the flowerbeds, the shrubs, the trees, among the stones. Now the trees have been uprooted and instead there are sealed, glistening surfaces like kitchen floors with narrow, inadequate drainage which prevent the seepage of excess rainwater, surfaces that contribute to the heat island effect and which will one day in the future contribute to inevitable flooding (to say nothing of the shocking crime against aesthetics). One day when I'm out running, I see a woman with a bucket and mop, swishing soapy water over, and *polishing,* her ugly driveway. I stare, glare and run on, a short, seething mass of pointless, muttering, ill-directed rage.

We make gardens to keep at bay the concrete, to ameliorate what we may see as the hard, bleak harshness of the urban world and then everything grows, expands, flourishes, even here; it encroaches and year by year, what we deem to be weeds appear in our paths and our gravel and our flowerbeds sprout with plants we didn't plant and our tree grows tall. It all takes time and work and we tell ourselves we're busy and don't have time, and so we pave our gardens and then cars stand where gardens did, as if we're subject and servant to their overweening metal will.

In this city, cars are all; large cars which take up space and use up oil (almost as if the people who extract it from the depths of the oceans feel the urgent requirement to use it up the most quickly). On any day, parades of huge vehicles pass, some of them bearing the names of Saharan nomadic peoples and Iranian pastoralists and I think about the naming of things, the calling of vehicles after the people whose way of life has been ruined by changes in climate, caused partly by the effects of driving such cars. Do the names, hinting at the perceived romantic freedom of nomadic life, gild the experience of driving a few children to school through a series of urban or suburban streets? (Might it be the case that the more inappropriately your vehicle is named, the less your sense of irony? This, clearly, is the inverse irony rule.) In an article in *Orion* magazine, Richard Louv writes of developers naming their developments after the species they have displaced.

On the first fine day after torrential August rain, a group of sparrows is bathing in a deep puddle in a quiet lane. I pass them when I'm on the way to the gym and stop awhile to watch their splashing, their eager flitting to and from the cover of an escallonia on a wall. If we lose sparrows, everything will change. Our lives will change, even if we don't at the time fully appreciate how. The loss will change the words we use, the ones we have available to us. It will alter our tenses, our measurement of years. They'll be lost before our eyes and as with every loss, our lives will be thinner, lesser; the future not only of the physical world but our mental world will be diminished, the world of our history and legend where the life of all our cultures resonates with all we've seen and all we've lived with, plant and animal, stone and cloud.

Without a species, we lose not only it but its history, its mirror in our own. We lose connections and while we may in the future read the words, we'll understand them less. A thread through time will be gone, another reminder of our long lives on this earth together. We'll have lost poetry if we don't appreciate (and smile at) Catullus's elegy for his girlfriend's sparrow:

... the sparrow, my lady's pet, whom she loved more than her very eyes; for honey-sweet he was ... Nor would he stir from her lap, hopping now here, now there, would still chirp to his mistress alone . . . Ah, poor little bird! All because of you my lady's darling eyes are heavy and red with weeping.

The sparrows continue their busy bathing as I pass carefully on the other side of the lane. It was, I remember, the Venerable Bede who was the first to mention the word 'sparrow' in English in a lovely musing on the nature of man's life in his *Ecclesiastical History*:

> The present life of men of earth . . . seems to me to be like this; as if, when you are sitting at dinner with your chiefs and ministers in wintertime, one of the sparrows from outside flew very quickly through the hall; as if it came in one door and very soon went out through another. In that actual time it is indoors it is not touched by the winter's storm yet the tiny period of calm is over in a moment, and having come out of the winter it soon returns to the dark winter and out of your sight. Man's life appears to be more or less like this; and of what may follow it, or what preceded it, we are absolutely ignorant.

On the radio, someone is talking of 'bunny huggers' and 'sandal wearers', and I wonder what kind of an attempt at an insult these terms really are? Sandal-wearing! I dream of sandal-wearing. I dream of something other than wellington boots or walking boots with thick socks and ice-grips, or, in particularly fine winter weather, shoes of a not too delicate kind.

As for 'bunny huggers', I remember with delight the rabbit who, for many years, lived between the house and the garden. David and Han had found her as an infant beside a dustbin on a country road. She wouldn't be persuaded back to the field from which they assumed she had come. They didn't want to leave her where she might be run over by a car, and so they brought her home. We bought her a hutch but never closed her in so that she ran free in the garden.

She came when she was called, asked to be let in, demanded to be let out, enjoyed the benefits of human company and the obvious joy of rabbit freedom. She loved people and she loved solitude, and demanded the opportunity to decide which she wanted and when. She was an astringent antidote to garden preciousness, and a reminder of the damage rabbits can do when let loose on edible things. I think of her dancing in snow, sending a fine, glistening spume spraying from her speeding, padded feet. She loved spring, kicked up her heels, running in wild circles on the grass. We all indulged in what is so scathingly described as 'bunny-hugging' (although we never applied so reductive a term to this clever, sybaritic, interesting lagomorph). The opportunity did not often arise. Hugging was, she made clear, a privilege and not a right.

July 28th

This afternoon, I pass the urban roost of rooks at the corner of Union Terrace Gardens, a sunken Victorian garden built underneath the soaring overhead viaducts, a continuation of the medieval 'Corbie Heugh' or 'Crow Glen', the defile through which the Denburn runs. The gardens are lovely; steep banks of shrubs and trees surround an area of grass where in warm weather, people gather to sit and sleep and let their children play in the sun. I've watched the rooks and their nests for 20 years and wonder how long they have been in continuous occupation of these trees, which tower over the busiest junction in the city. But now, there is a plan to level the gardens, to convert them to concrete and shops and car parks. The trees will go, the rooks' nests will go, Corbie Heugh and possibly hundreds of years of the residency of corvids will end. Where will they go? Where will people without gardens sit on the warm days of summer?

Memory is part of cities but not only for its human inhabitants. Others remember. In a fairly recent suburb built on farmland in the 1960s, rooks gather on the roofs of the shopping centre. They seem an unusual sight (although they do have favoured, long-established city-centre roosts). Crows and jackdaws congregate in the car parks of nearby supermarkets hopping among the cars, calling from the tree on the fringes of the buildings and I wonder if, in addition to the possible pickings, they are all there because their homes once were there.

In their splendid book *In the Company of Crows and Ravens*, John Marzluff and Tony Angell write of a parking lot at the University of Washington's football stadium where crows gather every morning, flying in from their nearby roost to gather on the ground. There seems to be nothing to draw them: no food, no water, no other apparent attractions. In spite of research into temperature and other possibilities, there is no explanation for this apparently social gathering. It transpires, when they discuss the phenomenon with a long-term resident of the town, that the parking lot was once a rubbish dump where crows gathered. In the light of this new knowledge, Marzluff and Angell suggest that the gathering is clearly a crow tradition, a cultural practice handed on from one generation of crows to the next, over some 40 years. Who knows what else they hand on, this genus of clever, big-brained birds, possibly conscious, possibly using sophisticated language that we cannot yet understand, the ways in which they carry stories, memories, creatures who have been social and organised for longer that we have ourselves?

I think of philopatry – site fidelity – of the devotion of creatures to and for their homes and birthplaces, to the sites to which they return after migration, their wintering grounds, the ones they may have dug or sought or built nests in for more time than we remember, places for which we have no respect and no caution, which we destroy in the furtherance of every interest of our own, in the casual way in which we remove habitats, cut down trees, destroying everything but the memory which is not destroyed in many of the species whose habitats they were.

We are all so alike. Some years ago, Edinburgh closed its old, overextended, unwieldy hospital. The vast spread of sprawling buildings of the Royal Infirmary, a labyrinthine Victorian chaos of forbidding wards, of towers and extensions and huts and crowded car parks, became obsolete and the hospital moved to new premises on the edge of the city. In the old infirmary, on the other side of Middle Meadow

Walk from the university where I studied, were the places where David trained as a doctor and the maternity unit where Bec was born, converted now to smart, expensive flats. New glass and metal blocks were built, and where the Accident and Emergency Department once was there is a supermarket and a smart café. Now when I walk past, I remember the lines of ambulances, the overhead lights, the bed rails glimpsed through windows, the nearness of other people's lives and deaths. But still, outside, on Middle Meadow Walk, there's often an assemblage of people sitting on benches, near where the entrance to A&E used to be. Most are unfortunates, people with memory but no place, people who drink from cans in the open air, drawn there because it's where they've always assembled, because of its proximity to a familiar place – the now empty department with the long-remembered comfort of the sort it's difficult even to imagine.

A talking crow has been encountered near Union Terrace Gardens. It greets David as he walks beside the granite balustrade on Union Bridge that separates the shopping crowds from the steep drop to the railway line below. A young crow walking on the parapet near him turns to him and says 'hello'. Anyone else might have been concerned for their sanity but David, having lived with articulate birds, knows the ways of corvids. He replies in kind. When I go next day to try to find the bird, it's not there. Instead, there's a man on the pavement next to the bridge, playing the bagpipes indifferently. Clearly it's a crow of some discernment.

And then, one morning, there's an amber alert. The website says, 'Aurora is likely to be visible'. This time, I contact people to tell them, all excitement and certainty. It might have been likely, but it isn't visible. There is on the designated night only a very faint glow of light in the sky. To compound the disappointment, the forecast suggests that soon the night of the Perseid showers, that brilliant display of falling meteors, will be a night of heavy rain.

The Fugitive in the Garden

On a chilly Saturday morning as I'm looking out of an upstairs window I see a small, furred creature processing alone along the pavement opposite the house. It's a red squirrel. I've never seen one in the city before. Some grey squirrels have a rufous tinge but this one, with its different shape, its different ears, I know to be a red squirrel, *Sciurus vulgaris*. He or she reaches the traffic lights while silently, I urge this lovely, much-to-be-welcomed creature to caution. I consider rushing out like some *Sciurus* lollipop lady to help it cross the road in safety but calmly and without mishap it does so of its own accord and disappears swiftly round the corner. I do not see him, or her, again. Before long, news reaches me that a red squirrel was indeed seen in the district, in the street adjacent to mine.

Squirrels are not unknown here. At least one used to be a daily sight in the garden, foraging, clambering up the branches of the *Hydrangea petiolaris* and thence into a bird-feeder shaped like a small house hanging in the arch of ivy; a way station on the approach to the fat balls squirrels appear to favour, a place from where it would swing wildly with a look I took to be alarm on its small, alert face. Even for a squirrel, there seemed to be a threshold of anxiety because soon it would

288

leap out of its dangerous refuge for the safety and stability of the flowerbed. I was never overly concerned about the possibility of any squirrel starving within the purlieus of my garden knowing that, with the resourcefulness and storing habits of squirrels, they'd always find food: handy tulip bulbs, nuts grabbed from the peanut feeder, things hidden and stored and dug up again from the flowerbeds. I stopped using the house-shaped feeder for bird food but left it hanging there all the same. It's a while since then. Every day, I'd see them approaching over the garden walls lightly, holding their magnificent tails like a fine antenna. But they were *Sciurus carlinensis*, grey squirrels.

I always welcomed them to the garden. Apart from their beauty and the charm of their manner and movements, they invoked the vague sense of gratitude I have towards most things wild that elect, *entirely voluntarily*, to spend time on territory I might loosely describe as mine. Only the necessity to deter rats encouraged me install the transparent plastic anti-squirrel dome that I did over the bird-feeders – squirrels are messy eaters – that, and a degree of regrettable parsimony since visits from Craig and Dod, while companionable and socially responsible, are expensive, as are the sacks of wild bird-food delivered with the gourmet seed mixes with which I continue to indulge the inhabitants of my retirement home for elderly doves. In spite of the plastic dome, the squirrels continued to feed, managing in imaginative ways to bypass my far-from-squirrel-proof system. What I always hoped for, unrealistically enough, was the agreeable level of mutual cooperation that would ensure that the squirrels left enough food for the birds. When they did come to the garden, I'd watch them forage and feed and then, every evening before dark, make an attempt to clear up the food scattered by both squirrel and bird. I'd watch too to see what I could observe of the squirrels' 'caching' behaviour – their food-storing activities. I have respect for food-cachers, those creatures such as squirrels and corvids who habitually hide things. Far from being a simple mechanism to store against future necessity, it's complex, involving stratagems and consideration of the behaviour of other creatures and a possible demonstration of the plasticity of neural structures, behaviour that certainly in corvids demonstrates both memory and 'theory of mind', the ability to appreciate others' mental states. I'd watch to see if any of this was evident in squirrels as they rooted around the places where, in autumn, I'd have planted next spring's bulbs.

Squirrels have about them an air of measured curiosity. I recall a day some years ago when I was summoned to the kitchen by a sudden, terrified outbreak of avian alarm-calling, a combined fright opera of starling, parrot, magpie and rook. Although I didn't see anything untoward – no cat in the garden, no visible hawk – the panic continued. It took a moment to find the cause: a small, furred triangle of grey pushed up against the glass at the corner where the rat-room roof abuts the kitchen window; an oddly frowning triangle between two flattened paws. Only when it began to move away was it identifiable as a squirrel which, with the half-alarmed, half-fascinated look of a small child at a zoo, had been peering in at the assembled company, nose and tiny paws pressed against the pane. Having observed us for a while – the undistinguished sight and sound of unbridled shrieking, wing-flapping, shameful, abject hiding and all – the squirrel dipped below the window and disappeared swiftly over the edge of the roof. I like to think that its look was the same one of bemused consternation that I've seen more than once on the faces of those contemplating my household, but probably it wasn't.

The squirrels have all but gone from the garden now but their going has, I believe, nothing to do with the mild obstacle of the anti-squirrel dome which succeeds in one purpose in that it appears to encourage many birds to the feeder that hangs underneath it, protecting them from the perennial rain. (In my continuing effort to deter rats, I lay a large piece of landscape fabric like a bird tablecloth under the feeder and regularly remove from it the residue of scattered seed.) The very few squirrels there are now feed instead, on their increasingly rare visits, from the wire dish I fill for ground-feeding birds. Even that much encouragement makes me feel as if I'm harbouring a fugitive.

I don't remember exactly when it was that I began to realise that tolerating and even feeding grey squirrels might be regarded by some as the actions of a subversive. I must have read something, come across an article in a newspaper perhaps or in a Scottish conservation magazine, telling me that grey squirrels are not to be encouraged. Some Scottish wildlife expert or other must have given an opinion, or a warning, informing me that grey squirrels must no longer be allowed to thrive within the borders of Scotland. I must have read that I should deny them food; that everything should be done to withhold from *Sciurus carlinensis* the means of preserving life. The reason for cooperation, I must have been told, is because these creatures are a danger to the population of *Sciurus vulgaris*, the 'native' red squirrel. Grey squirrels, I must have read, are 'vermin'; vermin that carry a disease both virulent and fatal which, being almost immune to it themselves, they're free to pass on to susceptible red squirrels. Greys, being originally American and introduced during the nineteenth century are, unlike reds, not indigenous to Scotland and only 'native' species must be encouraged, favoured, restored. Grey squirrels, apparently, are invading from the south, bringing this viral disease, parapox, a genus of the Chordopoxvirinae, with them. Someone must have told me that although I don't remember exactly who, or when.

From the first, it was the word 'invader' that distracted me. It took me to a world that extended past the squirrels in the garden, this word that resonated far beyond the lives and futures of all of us, including the family Sciuridae. 'Invasion' is, after all, a concept which can never have been far from the human mind, arising as it must have done with the demarcation of boundaries and the need for the settling of scores, consolidating resources at times of scarcity, strengthening throughout history as a possibility or a solution with every manifestation of competition and the desire for gain – for more territory, more water, more oil or even the unlikely, imponderable gains which, looking back on them, seem hardly worth the effort. The experience and threat of invasion remain the expensive reasons for standing armies and quantities of alarming hardware that works only because it's assumed or hoped, bewilderingly enough, that no one will actually ever use it. It is a double-sided concept of invaders and invaded that is portrayed, written about, the theme of our earliest literature, almost the first of our art; the development of war, its technology, ideas, morality. 'Invasion' is a

word that screams. It urges us to belligerence, to run or to repel. It con-
jures a world of borders breached, of unwarranted landings, of force
and change and imposition, of defilement and subjugation, everything
unknown and frightful, a phenomenon expressed in terms of tides or
swamps, armies or hordes, in the language of terror and of fear. It was
a word to ponder, one that made me want to measure how a word
might of itself alter our thinking, even about the fate of squirrels.

There are plenty of invasions undertaken in one way or another by
non-human organisms: by fungi, microbes, plants, creatures, by birds,
beetles, worms, potentially by everything that lives on earth (and by
humans too). Throughout the history of the evolution of species and
their distribution over the earth, organisms have travelled, moved on
changing land masses, on breaking, moving continents, used whatever
means available to establish their place on earth. Things fly, as they have
always done; they swim, cadge lifts and hitch and ride and hide on or
in every available medium, in ballast water and traded commodities, on
wind and rain, in plane, ship, car, on or in human or animal hosts;
thousands, millions of representatives of all the world's species moving
relentlessly over the surface of the globe. We are, after all, one space,
one flow of air and many of us are unstoppable and restless. Man too
has been responsible, both intentionally and otherwise, for displacing
and moving creatures from one place to another, by introductions and
travel and every other means available, the results of which are seen
only a long time later. What can be done about things living, thriving
in what's deemed to be the wrong place?

The field of 'invasion biology' has grown over the past 50 years, since the publication in 1958 of a book collected from a series of BBC radio talks given by Charles Elton entitled *The Ecology of Invasions by Plants and Animals,* a book regarded as the first to encompass topics previously seen as separate disciplines such as conservation, biology and biogeography as well as the possible future effects of historic introductions. Much as George Evelyn Hutchinson's 1959 paper 'Homage to Santa Rosalia' is considered to be the foundation for the future study of biodiversity, Elton's book is often seen as the point from which studies in the fields of ecology and 'invasion biology' began a timely expansion.

There are, it is clear, difficult, dangerous and almost intractable problems in the introduction and spread of species. There are implications for health, for economies, for ecosystems. Plants outgrow others, threaten crops, destroy the habitats and lives of pre-existing species. Pathogens spread, mutate and may endanger the entire world but in dealing with these species, out of place and time, what do we do and how do we do it?

If the subject, scope and magnitude of the problems of 'invasion biology' are complex, the terminology is no less so. What is indigenous? Is indigenous the correct term? What about 'non-native', 'introduced', 'alien', 'exotic', 'invader'?

In the introduction to his excellent, thoughtful book *Invasion Biology* published in 2009, Mark A. Davis expresses regret that his field of study ever began to use the terminology of warfare, pointing out that, while he himself dislikes the term 'invasion', he can hardly write a book on the subject without using it. He says too that the problems of using military language are that, while it might 'attract a group of highly motivated supporters', it can also make conflicts more difficult to resolve. Davis presents the problems, benefits and analyses of possible approaches to problems posed by 'invasive' species, as well as examining some of the underlying questions of political, psychological and philosophical motivations and effects in dealing with them.

In a time of change, of disappearances, extinctions, losses of species and habitats happening at an accelerated rate, the importance of knowing what is where, and whether or not, if it is there, its presence is beneficial or otherwise, is crucial. If plants, creatures and organisms of one sort or another are threatening lives, habitats and livelihoods, clearly it's as well to identify them and to take what action is possible to limit any damage, to restore and to reverse, to protect, wherever such protection is practical or possible. There are many organisations devoted to defending the world against the dangers of invasive species and many of them publish lists of the endangered, the critically endangered, the lost, the disappeared, the most dangerous, the most undesirable, the 100 worst species, the 100 most invasive. There are databases and acronyms, slogans and exhortations – everything, in fact, to draw our attention to the dangers posed by the wrong things in the wrong places. (There does seem to be a certain zeal that accompanies the endeavour. One organisation devoted to the extirpation of invasive species has a snake, fangs extended, peering threateningly round the frame of its home page. Another has a banner: Let's Stamp Out Invasive Species!) We all know of giant hogweed, of Japanese knotweed, New Zealand flatworms, ruddy ducks and Chinese mitten crabs, all of which have found their way (some invited, some not) into our environment. Among the species I find on a list of the world's worst invasive species are:

Canis lupus: grey wolf (it does not say if *Canis lupus familiaris*, the domestic dog, is included)
Capra hircus: domesticated goat
Felis catus: cat
Columba livia: common pigeon
Rattus rattus: black rat
Rattus norvegicus: brown rat
Salmo trutta: brown trout
Scurius carlinensis: grey squirrel
Vulpes vulpes: fox
Mus musculus: mouse
Sturnus vulgaris: starling
Oryctolagus cuniculus: rabbit

In this list, perhaps, is the essence of the nature and extent of the problem of the science of invasion biology. What, after all, are we to do about these species? Even if we could do something, at which point was anything in the natural world fixed and immutable? To which year would we return, to which golden age or stage of the life of earth? The Pleistocene? The Eocene? Which window of past perfection might we attempt to restore, or replicate? To which moment might we wish to retreat, pull up our drawbridge, erase from memory what we ourselves,

296

or the processes of nature and time, have wrought? Species change, travel, become extinct. Aurochs, wolf, elk and lynx were once native species in Britain. (My own choice, should anyone ask me, is for a speedy return to the Cretaceous, that period of vast and mass expansion of species of all sorts, when in mild and encouraging climates thousands of bird species evolved.)

Reading these lists, I reflect on how easy it might be to become fearful and phobic. From one, I learn that I have at least one of the most undesirable – a *Rosa rugosa* – in my garden. ('The Rugosa', my favourite old rose catalogue tells me, 'is a native of Japan and was brought to Europe in 1779. Ideal subjects for poor conditions. *Remontant.*') What shall I do? Shall I wrench it asunder, burn it, dig out the soil, sterilise the place where it has grown, or shall I wait until June for it to flower into perfect whiteness again, this fine-petalled rose called 'Blanc Double de Coubert' which grows by my front gate, a rose with a scent so sweet and captivating as to stop passing strangers? 'What is that?' they'll ask as they stroll by on a rare mild summer's evening, transported as I am by the scent to another time and another world when roses had scents and possibly, lamentably, we weren't much interested in whether or not species were dangerous invaders from another place. (Communication between lists, or their compilers, is clearly limited. One useful remedy suggested for dealing with this unwelcome, rampant foreigner of the rose variety is to employ the destructive services of the equally invasive goat.)

While some species are dangerous and damaging, the perceptions of harm done by other types of 'alien species' are often incorrect.

Himalayan balsam, that perceived scourge, that plant of menace and 'invasion' is, in fact, a plant misjudged. In his paper, 'Plant Introductions' published in *Silent Summer*, Andrew Lack suggests that there is no evidence 'that any rarity has declined as a result of its spread'. He says too that as the preferred plant of declining bumblebees, its presence might be welcomed. But still, parties of worthy Himalayan balsam-fighters spend weekends waging war against an enemy that is, perhaps, no enemy at all.

Observation of the phenomenon is not new. Pliny the Elder writes of introduced species, of balsam and cherry trees, of vines and bamboo. Charles Darwin, with the exacting and detailed thoroughness he applied to the study of worms and everything else, followed the possible routes and spread of seeds, examining them in bird faeces and stomach contents, on logs, in mud, on disembodied avian feet. He experimented to discover if seeds would germinate after long immersion in salt water, and if seeds from the crops of long-immersed birds would, too.

In Aberdeen, in the latter years of the nineteenth century and well into the twentieth, the professor of botany James Trail kept vigilant watch on the exotic species being introduced by rail and sea at a time of increasing foreign trade. In his comprehensive account *The Flora of Aberdeen* Professor Trail describes his findings from the harbour and rail-side sites he monitored closely for decades. He lists grapevines, date palms, cannabis, lizards brought in with esparto grass, Canadian pondweed and blue lupin, just a few of an amazing 470 species of alien plants he found, some imported with ballast water but most brought mainly from Europe and the Black Sea with grain.

Some of these may have been brought in commercial traffic of other kinds e.g. the bur-like fruits of some medicks, of burdocks and of Xanthium are entangled in wool, others in moss-litter and so on. Others e.g. fruits of hemp and of canary grass, are largely dispersed because of their use as food of cage birds.

(The place where Trail did most of his collecting and watching is now covered by a large new shopping centre. Even the grim old coal yards behind the railway sidings which first subsumed Trail's hunting ground are covered over by plazas, chain restaurants, shops and cars.)

The Victorian passion for collecting has had some interesting results, from near extinctions of several species of birds to the introduction of some of the invasive species we're still trying to deal with. Rhododendrons in a nearby garden, once the home of a Victorian plant collector, part of his unique and extensive collection, were brought back from journeys to the Himalayas and other parts of Asia at a time when collecting, the expansion of knowledge and plunder were inseparable, when the most invasive of all species (as it still is) was man. His rare and carefully collected specimens were and are invasive; the rhododendron a species that has spread widely and ineradicably.

One day I read in a newspaper that a local wildlife organisation is supplying traps to deal with the vexatious presence of *Sciurus carlinensis*. I phone to enquire about the details. A young woman answers, speaking with a degree of undisguised, ebullient pride as she informs me that across the city, *more than twenty people* have already availed themselves of this most valuable service. I ask what has to be done

when an alien invader is caught. I must, she says, inform them immediately and someone will arrive speedily to 'dispatch' the prisoner.
Would I be interested? I tell her that I wouldn't.

After my call, I read more about the culling programme. The evidence, as all evidence, is subject to question. Every aspect, when I look,
is disputed; all of it is difficult, if not impossible, to prove definitively.
The length of the duration of the red squirrel's tenure in Scotland is far
from certain. There is no reliable evidence that it has been here for
thousands of years, one of the criteria for its being regarded as a 'native'
species. There is no record of red squirrels in Britain before the fifteenth
century and following a decline in numbers, red squirrels were reintroduced to Scotland from Scandinavia. Red squirrel populations have
been declining for hundreds of years and, in common with most other
species suffering a reduction in numbers, the principal cause is human
activity – the degradation of habitats, persecutions and changes in climate. But does this matter? For squirrel red or grey, or for me, does it
matter how long our separate roots have extended down into the
doughty soil of this particular nation? Not that long in squirrel terms,
or fossil ones, but in human terms? For long enough? Not quite long
enough? Who decides?

I read too the evidence about the spread of squirrel parapox virus,
which may have been introduced with the grey squirrel at the end of
the nineteenth century. I read of the epidemiology of the disease, the
role of fleas, the symptoms, the ulcerative and exudative dermatitis, the
epidermal necrosis, lesions and scabs to which red squirrels are particularly prey; of possible immunity and the development of vaccines and

when I've finished, I wonder what difference this all might make to my consideration of the fate of squirrels. Grey squirrels have developed immunity to parapox virus, to which they were once susceptible. Evidence suggests that red squirrels may be developing immunity too. I read about the efficacy, or otherwise, of culls, and of species which have both survived and thrived following attempts at culling.

It's not only the facts that have been presented for implementing a squirrel cull that catch my notice. A scientist involved in the culling programme writes in an article of an indefinable quality possessed by red squirrels. I read of the special beauty of the creatures, their 'charisma', their 'iconic' status. I read of the tourist potential of red squirrels and of the 'natural affinity' that exists between Scots and red squirrels. *Affinity. Natural.* The words suggest something beyond mere liking, something possibly beyond argument itself, except that nothing is. (Natural? If not, unnatural.) More than one dictionary definition of affinity includes worrying, indefinable concepts such as spiritual attraction. I don't really feel that I have 'an affinity' with red squirrels. I like red squirrels. I would like to see them preserved and thriving. Is this the same? No affinity? Maybe this is an indication of the degree, the measure of my or anything else's Scottishness.

I might be thought to have an affinity with corvids but it's other than that. I like and admire them. I envy their abilities as a genus. Among their numbers are ones I love and have loved. I enjoy the company of the ones whose lives I share but as for affinity, how does it compare with what appears to me still to be the close and mutual, possibly commensal, relationship I have with the limited number I know well? Am I missing some subtler aspect of the term? Is there a spiritual aspect to our relationship? Not that I've discerned, or not on my part at least because, even after 25 years together, Chicken's spiritual leanings remain opaque. Mystical connections? I wish.

This assertion of 'affinity', however, forgets history. For how long has this 'affinity' existed? When did the 'indefinable quality' make itself known? Were they both extant when the killing of red squirrels was common in Scotland? In his book *Silent Fields*, Roger Lovegrove refers to accounts suggesting that the red squirrel, on the verge of extinction in the Highlands in the seventeenth century, was subsequently reintroduced, squirrel numbers then growing to the point where it was seen as a pest and 'shot in enormous numbers'. Lovegrove writes of the Highland Squirrel Club established in 1903, which killed 82,000 red squirrels over thirty years. Between 1818 and 1890, 13,875 red squirrels were killed on the Strathspey Estate alone.

The author of *Scottish Mammals*, Robin Hull, writing in an online archive, talks of the deforestation that was responsible for the near elimination of red squirrels from Scotland in the eighteenth century and of the report from the Old Statistical Account of Scotland of the rarity and even possible extinction of red squirrels from the Highlands. He quotes from the report: 'they are destructive of the small birds by devouring their eggs and are extremely injurious to young planting by barking the tender shoots'. Hull writes of the distinguished Scottish naturalist Frank Fraser Darling describing red squirrels as forestry pests whose presence was seriously injurious to birds, and of reports of numbers of red squirrels killed by the Ross-shire squirrel club: 3,988 in

302

1904, 739 fewer than the previous year. Bounties were awarded on the presentation of tails but, Hull suggests, the more enterprising cut off the tails and allowed the squirrels to go free, a certain way of producing more tails. Affinity?

(The Statistical Accounts of Scotland are themselves fascinating documents. Published in the 1790s and 1830s respectively, the Old and New Statistical Accounts were compiled, parish by parish, often by the minister, doctor or schoolteacher, and present detailed contemporary pictures of every aspect of Scottish life.)

In his paper 'Concluding Remarks' given at a biology symposium at Cold Harbor in 1957, eminent ecologist G. Evelyn Hutchinson, in discussing the role of niches in ecology, cites the view of the charming red squirrel being displaced by the 'bad, bold invader' as mythological. The truth, Hutchinson said, was that both species were persecuted by humans and that the grey squirrel, being more able to withstand persecution, spread into areas from which red squirrels had retreated.

As I read what I can on the subject, the search takes me inevitably to Internet forums, in particular one linked to a Scottish wildlife organisation. I should be used to the tenor of web forums, but clearly I'm not. What shocks me is the intemperance, the language used. It is being used of animals but is the language of extreme nationalism, of violence and vengeance and hatred. ('Incursions from England by diseased grey squirrels . . .') Red squirrels are Scots and greys are not. It is as if the destruction of foreigners is paramount. With the brutal, unconsidered ease of Internet exchange, people possibly gentler, more humane in reality than they appear in the chill glassy square in front of me may be turned demagogue or would-be murderer by proxy, under the cover of these anonymous freedoms.

On a cold day in spring, I watch a red squirrel for a long time as it feeds on the other side of a window at a hide in the Kielder Forest in Northumberland. It still has the long ear tufts of winter, its fur a rich, deep-terracotta red. It is as lovely, as delicate a creature as it would have been while it was being persecuted as a pest. As I watch it, it occurs to me that all that changes is human perception.

The evolutionary biologist, the late Stephen Jay Gould, a lifelong critic of 'sociobiology' and biological determinism – ideas suggesting evolution and biology as the principal elements in the formation of human and social behaviour – discusses in his paper 'An Evolutionary Perspective on the Concept of Native Plants' the beliefs and misconceptions relating to the benefits and superiority of 'native' species of plant and of their consequences. Trenchantly critical of the view that 'native' species are necessarily optimally suited to their environment as a result of the processes of evolution and are therefore to be preferred, Gould traces the roots of this thinking to the mistaken idea that evolution works towards perfection, much as pre-Darwinian religious thinking held that in a God-created universe, every aspect of the natural world was designed to be in its perfect place. The results of this thinking, he suggests, encouraged the ideas adopted by some of the most egregious thinkers of the twentieth century in their associations of the 'purity' of plant species with the racial make-up of the populace.

'Native' species, he suggests, are not the species best suited to a particular habitat: they are the species that got there first and managed to survive and their having done so suggests nothing about their superiority or suitability over competitor species, and in trying to suggest otherwise lies the danger of claiming 'that my "native" is best and yours fit only for extirpation'.

It's some time after my phone call that I read that the organisation which supplies squirrel traps has instituted a scheme whereby one can text them to report the sighting of a grey squirrel. The sightings are to be plotted, the pattern of grey-squirrel occurrence noted and action ultimately taken against them. I shiver.

Over the months, the years, I continue to watch for squirrels but see fewer and fewer. By now, the garden is almost squirrel-less. I rarely see any in the city's parks or in its trees. I wonder if the cages in other people's gardens may be filling up; if vanloads of squirrel exterminators may be rushing through the streets to carry out their duties. I hope, vainly and foolishly, that the grey squirrels have of their own volition gone elsewhere. I read that the red squirrel is appearing on the fringes of the city at Hazlehead Park, a mile or so away. And then I see the red squirrel across the road and I'm delighted that it's there but would have directed the same pointless, silent warning about traffic and caution towards a grey squirrel. I would have welcomed it equally, and when, not long after that, I see one running in its freeze-frame, stop-start way across the grass in Queens Terrace Gardens, I wonder, even as I watch it, if someone is texting to alert the appropriate authorities that it's there. I wonder for which species, which individual of my own or any other species, might I do this at someone else's request? Which are the words, the ideas that might make me comply?

Then, on a morning in late summer, after cleaning and filling the containers hanging from the bird-feeder, after replacing the water and ensuring that the ground-feeding birds are provided for, I look from the window to see a grey squirrel sitting in one of the containers. It fits neatly into the round wire feeder dish. It is months since I saw one. I watch it with a sense of welcome and anxiety. It is there for a while, holding up small pieces of food delicately between its neat and perfect paws. I have only to turn my head as I'm at my desk to watch the frantic, dark-eyed diner. I watch it with an altered sense, admiring its beauty in a different way, as a fellow inhabitant of fragile space and time.

Roger Deakin writes of squirrels being the only wild mammals to come to gardens and parks, and of their removal as an indescribable loss. They move so lightly. They are more measured than any other creature; you hold your breath with their sudden stillness, their total immobility, the frozen comma of the fur of their tail, their watchful eyes.

Looking at the creature, I think about the trapping and killing of squirrels itself, and know that the arguments for and against the squirrel campaign are only part of the point. The other is the power of what we say. I read a headline in a Scottish newspaper: 'Asian Invader in Scotland as Ladybird Numbers Drop', and wonder if the person who wrote it believes that we can segregate ideas and consequences. Another piece, relating to the economic price to the country of 'alien animals', details the guilty parties and the expense they are supposed to incur in a box headed: 'You're not from round here, are you?'

I think again of John Locke's observation that 'the killing of beasts, will by degrees, harden their minds even towards men'.

Days later, I see something high in the branches of the large cherry tree in the next-door garden. I think it's an owl but it's not. Through binoculars, I see that it's a grey squirrel, apparently asleep, huddled in the rain on the top branch of the tree and as I watch it, I think of it all again; of how, as humans, we feel free to treat others, both human and not, and of the ways we find to explain to ourselves what we do, the ways in which we present and elevate our condemnations or our judgements. I think of the dangers of separating the lives of animals and men, believing them to be irreconcilable or subject to different standards; of assuming righteousness, of the consequences of words once spoken that we can never recall.

INTO AUTUMN

August 10th

August now; the turn of the month from high summer, and it feels already as if it is a step towards autumn, an interim but a decisive one when we know that if there hasn't yet been a true summer, there won't be now. The few warm days of July were all we'll have to mark it, one brief flare to keep an entire season in our minds. The sense of shutting down, the inexorable path from solstice to equinox has begun earlier than usual with the experience of waking to cloud and rain and almost darkness, wondering for a moment where exactly in the year we are. There's a sense of unease about a season incomplete or unexpected, a necessary readjustment of expectations and of one's own place in the scheme of time.

In the early morning, I walk through the town gardens in front of one of the Victorian terraces of large imposing houses, now turned into offices, past crows pottering on the grass, towards the station and the train to Glasgow. There's a fierce wind, a pale sun, the haze of rain clouds spreading in a strangely livid lilac sky. The rail line south runs by the harbour and then out along the top of the cliffs. Most of the harbour cetaceans will be heading for deeper water soon, the white-beaked dolphins and minke whales, the harbour porpoises. The bottlenose dolphins will be back in numbers to spend the winter.

We travel high above a North Sea that's tumbling, etched steely grey and white. In Angus and Perthshire, fields have half disappeared under sheets of sky-glinting flood. Every pond and lochan's full, the rivers too, the Tay at Perth almost still and brimming. All day, there is rain; in Glasgow it is raining as, in memory at least, it always is. When I get back to Aberdeen in the late afternoon, I come back to another season. The gardens I walked through in the morning are strewn with newly fallen branches, dark and soaked from heavy rain.

Today is the anniversary of Ziki's arrival, four years since a very small, nervous, silent bird came to us. He's still nervous, but no longer small or silent. He has, by means I can't quite understand, acquired a full range of crow calls. He shouldn't have. He is one of the birds whose song is learned but I don't know where he learned it. For all its mystery, his calls are loud and expressive. He has changed, progressed slowly. He plays with the many bird toys I buy for him. He has made his own cache site in the wall next to the back door. He has a collection of semi-precious stones, turquoise and amethyst, pink quartz and chrysoprase. He has favourite colours and I find his choice on the floor of his house: the chip of topaz stone, the piece of sea glass, the translucent blue bead. He carries in his beak a grey rubber mouse. He is messy. Only occasionally do I think about the organisation of the room where he lives, of the day when it was bird-free, the surfaces white or wooden and shining gently, when the tile floor was scoured and unencumbered, when I could walk freely from kitchen door, past washing machine, sink, tumble drier (used now only infrequently in a very small endeavour to save the drowning world), past food-storage cupboards and odds-and-sods cupboards, past wine racks and floor buckets and brooms, all dusted and swept and tidied away behind doors in minimalist appearance of almost emptiness, to the back door which I could open without thought, leaving it ajar to allow the douce seasonal breeze (or more usually, the damp, chill air or freezing rain) gently to enter the long room and penetrate, with purifying effect, the far reaches of the rest of the house where together, Chicken and I would smugly contemplate the perfection of our orderly domestic universe. It's but a vision, almost forgotten. I look at the entry on the calendar: 'Collect Ziki', and it's without regret that I reflect on changed circumstance.

August 11th

The city is quiet today. It's as though it has shut down for the duration (a bit like Paris does in August). Sensible people are still far away from the chill of this late summer, away from today's light dusting of rain, which seems the only alternative provided to heavy, drenching rain.

Already, another season of *Zugunruhe*, the restless urge that impels migration. I see it in the swifts who are already leaving. There are some still flitting, shrieking, looking few and lost against a watery grey sky. Others birds will be returning or preparing to return, passing over, stopping by, visitors breaking the journey from breeding grounds in the High Arctic to their wintering in sub-Saharan Africa, stopping briefly from this mysterious realm above our own, where these indomitable journeys are under way, a cross-continental, transglobal airway system of complexity and indefatigability that beggars our own.

But I'm still here, busy with work, and so is the city's population of gulls, busy with the rearing of this season's young. On the playing field of a school near the centre of town, a collection of infant gulls is resting on the grass, twenty or so of them, their parents lying nearby or padding around them in the careful way they do. A row of soft-feathered, grey-brown chicks stands neatly along the edge of the roof of an old garage on the periphery of the field, lined up one by one on the moss-covered grooves. The schoolchildren are on holiday, and although their usual place has been taken over as a gull crèche, the holidays are ending and term will begin again soon. I watch them for a while, this peaceable domestic scene, adults and chicks dotting the grass, pursuing their social, complicated lives.

This is gull season, even more than the other seasons in this sea-edge town of harbour and fish-houses and long-term city-gull residences. All year round, gulls and their young are everywhere, feeding, flying, calling. The sight and sound of them are as much a part of the place as the stone or air but now is the culmination of the months of gull preparation for nesting, egg-laying and hatching, months during which there have been calls from rooftops, imperious white heads peering from among chimneys; weeks when the fledglings were being fed in the nest.

Although the gulls' move inland has been fairly recent in other places, Aberdeen's gulls are long-time residents, nesting on city roofs for the past fifty years at least. Among the reasons we know of for gulls becoming more urban are the usual ones, the ones we caused ourselves. Following the introduction of the Clean Air Act of 1956 prohibiting the burning of waste, gulls began to feed at landfill sites. In the years since, as we've increasingly depleted the life of the seas and filled the streets of our towns and cities with edible rubbish of every sort, gulls have moved into cities to feed on still more of the things we've left behind.

Most commonly, the gulls are herring gulls, *Larus argentatus*, the birds referred to dismissively, often contemptuously, together with every other large, white-and-grey coastal bird as 'seagulls'. (Many gulls, such as black-headed gulls, aren't 'seabirds' at all but spend most of their time inland.)

When I think about the word and the way it's used, I regret that so many people in urban areas where gull numbers are increasing appear to have so little time or sympathy for these remarkable birds. It seems ironic too that a single word *seagull* should be used of one of the most complex families of birds to be found on earth; birds whose phylogeny, taxonomy, identification and distribution are subject to the finest scrutiny and passionate discourse, to constant reclassification and re-evaluation, whose every aspect causes extended scientific debate and continuing uncertainty as to who and what exactly they are, and where they may or may not fit into the noble roster of avian species.

Mainly northern birds, *Larus argentatus*, of the order Charadriiformes and family Laridae were thought for a long time to be part of a 'cline', a ring of gull species extending round the circumpolar north, species of sufficient genetic similarity that they were capable of interbreeding with the species geographically nearest to them but which, when they reached the end of the 'ring', could not breed with the species at the other end – *Larus argentatus* at one end, *Larus fuscus*, the lesser black-backed gull, at the other, with a chain of six other *Larus* gulls in between.

The concept of the 'ring species' came from a study carried out by the late Ernst Mayr, one of the world's most distinguished evolutionary biologists who died in 2005 at the age of 100. Writing in his book *Systematics and the Origin of Species* in 1942, Mayr suggested that 'speciation', the process by which new species arise, had occurred with the geographical movement of specific gull species to the circumpolar region where, given the circumstances imposed by distance, they evolved into a 'cline'.

It was in 2004, with the publication of a paper in the *Proceedings of the Royal Society of London*, 'The Herring Gull Complex is not a Ring Species' by Liebers et al, that the 'ring' theory was disproved. Studies of the mitochondrial DNA of 21 gull species demonstrated that previous beliefs about the descent and relationships between gull species were unsubstantiated and the cline theory wrong. (Delightfully enough, proving an encomium given to him by Stephen Jay Gould who described Mayr as 'the greatest taxonomist of the twentieth (and twenty-first) century', one who taught the idea that taxonomies are 'active theories' rather than ones which fix knowledge in time or place, Ernst Mayr was apparently thrilled by these new research findings and shortly before his death at the age of 100 wrote to the researchers, in his own hand, full of enthusiasm and praise for the work that had overturned his own.)

Attempting to follow recent reclassifications and renamings of gulls is dizzyingly complicated, the only wonder (and joy) being that there are people devoted to this study, for whom it matters hugely that *Larus cachinnans* and *Larus michahellis* are recognised as distinct species, or that gulls should be appropriately assigned on a phylogenetic tree according to which, the last time I looked, there were 28 species of *Larus* gulls alone, to say nothing of the genera Chroicocephalus, Leucophaeus or Ichthyaetus, who have another twenty or so between them, and all the others too.

Visible, audible, omnipresent, drifting endlessly in the sky above us, *L. argentatus* is another of the 'urban exploiters' who seem numerous, safe in their very existence, but who aren't; birds whose numbers have been falling until now they occupy their own place on the 'red list' of endangered species in the UK.

Just as Konrad Lorenz's name will be for ever linked to jackdaws, so the name of Niko Tinbergen's will be to herring gulls. (Lorenz and Tinbergen shared the 1973 Nobel Prize for Physiology or Medicine with Karl von Frisch, the zoologist who was first to discover the 'waggle dance', the mode of information transmission in bees.)

In 1953, Tinbergen, a Dutch zoologist, published what is still the most authoritative study of *L. argentatus, The Herring Gull's World: A Study of the Social Behaviour of Birds*, a comprehensive, detailed examination of the lives and behaviour of the herring gull, one of the foundational documents of ethology, the study of animal behaviour. It was a lifetime's work:

'Having had the good fortune to spend my boyhood and much of my later life in Holland by the sandy shores of the North Sea, I have naturally come under the spell of its most familiar bird, the sturdy yet graceful herring gull.'

Few people who read this book could continue to see these birds as little more than pests. Writing of their physical, mental, social and emotional lives, Tinbergen details every aspect of the lives of herring gulls, every detail of their behaviour, senses, their breeding, the rearing of young, their complex social relations, their angers and fidelities, their dogged efforts just to live.

August seems late in the year to see these young birds but gulls take time over the business of breeding, from early spring when they begin the finding, or re-finding, of their mates. Monogamous, capable of mutual recognition, of respecting their neighbours, living amicably with their partners, gulls are faithful to their homes, practise 'site fidelity' and return to the same nest sites annually. (Mated pairs may have spent the winter apart but will reunite for breeding, both returning to the same site, recognising their partner from some distance on his or her return.) The prolonged process unfolds amid the choosing of nest sites, the construction of nests, the laying of eggs in mid-May. Then there are four weeks of incubation and a further six weeks while the chicks are still in the nest so that only now, in August, the carefully nurtured young are out and about, with the prospect of being fed by their parents for a further three or four months. Gull families gather on pavements. Young gulls stagger, bend, beg, scream at every corner, wander onto roads, stand helpless amid traffic, cause heart-stopping anxiety to drivers and watchers alike. (They need crossings, or safe areas or better, no traffic at all. It's yet another of those times of year when again you realise how little you can do to intervene in the lives of other species.) Along the lanes, in the car parks, long-legged brown and white chicks are easing themselves into the lives they will lead, if they're fortunate, for the next forty years or so.

In the *The Herring Gull's World*, Tinbergen details every aspect of gull behaviour: the nature of their territoriality, how they have evolved their complex methods of warning and negotiation, their pair formation and courtship diplays, their rearing of young. Perhaps the best known aspect of his work is his study into the function of the red spot on the lower mandible of the adult gull. Chicks respond strongly to the stimulus of the red spot, aiming towards it in begging to be fed. In experiments, Tinbergen showed that the colour and contrast between the bill and the spot is what attracts the chicks, and that the precise shape and angle of the parent's bill is important in initiating feeding. Even the crudest wooden representations of a gull's head will evoke the begging response as long as a red spot is painted onto it.

Sternly critical of anthropomorphism, Tinbergen writes nonetheless with great tenderness of the subjects of his studies:

> With the hatching of the chicks, a most charming period begins ... They were not at all shy ... as long as they were kept in a nice warm spot, they would not defecate. But when they were handled, there were accidents. Sometimes we had them run from one end of the tent to the other just by calling them in turn. They were fascinating little creatures.

In a newspaper article, someone complains of the sound of gulls in London, on the grounds that they 'squawk'. 'Squawk'? Tinbergen writes: 'The voice of the herring gull is wonderfully melodious ...' and I agree, it is. There are few sounds as evocative, as stirring as the profound, plaintive beauty of their calls. Tinbergen enumerates the calls of *Larus argentatus*: call-note, charge call, trumpeting call, mew call, alarm call, 'choking' and the sounds made during courtship and mating. The 'mew' call, the one most associated with desolation and *tristesse*, the call that seems to be the summation of loneliness and sorrow is, Tinbergen says, nothing at all to do with sadness, but instead 'indicates breeding activity with an emphasis on the friendly attitude towards mate, territory, nest and young'. Having always enjoyed the sense of plangency in their calls, I don't know whether to be pleased or disappointed. (I'm glad of the sound of gulls now that the swifts are leaving, taking the essence and savour of summer with them. The oystercatchers have gone too although I still listen for them every evening at the moment when there should be that high and joyous calling.)

Walking home from town, I pass the abandoned garden, flourishing now with the abundant rain of summer. It's not exactly a secret garden because it's quite visible from the lane that runs behind a handsome row of granite terraces near the centre of town, the garden of what was once a house, converted now to an office. The back wall's broken down, a branch of espaliered apples hangs neglected against the uneven stones, a cascade of dark, fine-leaved ivy has sent exploring fingers to root into the broken edges. Gardens here revert with neglect as they do everywhere, not quickly in a single season as they do in warmer climates, where everything becomes overgrown almost before your eyes, but inexorably, slowly, over years. The once neat planting has been infiltrated by

320

broom and dock, the fuchsia and hypericum grow unchecked. A self-seeded yew is growing by the wall. The branches of a weeping pear are untrimmed, its silver and jade leaves sweeping to the ground almost covering, half hidden, a climbing rose. As I pass, there's calling from the roof of the granite terrace as parents and young gulls engage in a sound-exchange of request and feeding. I always know I'll hear them here because this is the same gull-nesting site that was described in a book about the wildlife of the city published 30 years ago.

Larus argentatus are the gulls who share our roofs and streets, but south of the harbour, and at places all along the coast, there are the kinds of gulls and seabirds people might choose or make an effort to see, the birds that, till now, are content to spend most of their lives at sea or beside coasts and cliffs. At the edge of the city, near the harbour, there may be, at different times of year and weather, great and lesser black-backed gulls (the former larger, darker), the misnamed common gulls, which are not very common at all, the large Arctic glaucous gull, the smaller kittiwakes with yellow beak and black legs, the yellow-headed gannets, the low-winged skuas, the sea-dwelling fulmars, related to the albatross, or the Manx shearwaters, birds unable to walk on land at all. Often there'll be cormorants in the bay, slim and dark and holding out their wings, and shags with their crested heads visible among the eiders and oystercatchers and mallards. (Stand in the wind above the harbour and the city and they may pass overhead, on their way somewhere else, these crossers of distances beyond imagining, these creatures related to our terraqueous world but belonging to a universe of their own, one of little but ocean, sky and flight.)

I look but know that with most gulls, I'll probably never be certain exactly which any of them are. Gulls confuse; they defy attempts at easy identification. If the difficulties of naming gulls and of sorting out their relationships, species and subspecies appear legion, they're nothing by comparison with identifying gulls. The trouble is that to the untrained (and often the trained) eye, many gulls look quite alike, clothed as they usually are in discreet plumages of white, grey, black and brown. There are many morphological differences in the shape of body, beak, feet, legs and eyes, to say nothing of the endless possibilities provided by feathers, but even if you've mastered those, there are still the patterns of gulls moulting, of three- or four-year moults, the fact that it can take years for a gull to reach its definitive plumage, passing through phases of natal down, pre-juvenile moult and juvenile moult and that there may be only very fine differences between species. Added to this are the endless variants and hybrids, and the effects of weather and light on the day you're exercising your perfect eyesight and heroic devotion to an apparently labyrinthine, never-ending task.

August 22nd

I'm on my way to the first day of the conference on conflicts in conservation. All around me, people clutching papers and laptop cases are making their way to the Arts Centre where it's being held, with an air of earnest excitement unusual for Aberdeen on a Monday morning. As I approach the pillared portico I notice, at the foot of the steps, a herring gull standing, engaged in dialogue with someone out of sight. When I turn the corner, I see a young woman who is sitting on the steps, sharing breakfast with the gull, chatting to it. The gull graciously accepts the large pieces of bread she extracts from a paper bag. It's not a sight you see here often, the purposeful feeding of *Larus argentatus*. Mostly, there are complaints about their nesting noisily, messily on roofs, picking at rubbish bags and bins, about the snatching of food from people's hands. (Gulls are considerable omnivores – Tinbergen's descriptions of their diet makes unnerving reading but we already know from observation that they'll eat the most horrible of fast food – some of the food really is fast: they'll eat other birds and even rabbits.) The gull expert Peter Rock suggests that food-stealing is learned behaviour in gulls, and his very sensible answer to the complaint is, 'Protect your food.'

This conference is a conference of experts. Over the next four days, some of the world's most distinguished thinkers on the subjects of ecology and biodiversity will discuss how best we all may live together on this planet. The topics will span the world, from conflicts over land use in Hungary or southern Africa to the management of snow-leopards or leatherback turtles. We will hear papers on bluefin tuna, eagles, deer and beavers, on peace studies and anthropology and more; a bewildering, dazzling display of knowledge and research.

Later in the day, I see the young woman who was feeding the gull sitting in the cafeteria with a group of colleagues, all from South America, all environmental scientists of one sort or another. I wonder if the gull was unfamiliar to her, if she was simply interested in a new species or if her profession inclines her towards acceptance and mutuality. Does it take unfamiliar eyes to value what we have?

On the way home, I stop for a moment in front of one of the city's architectural treasures, recently restored, the splendid granite facade of Marischal College. From the scruffy, dark and dirty brooding span of building that it was (the second largest granite facade in the world after El Escorial, that vast, austere sixteenth-century complex of royal buildings north of Madrid – a fact to be reckoned with), after long and patient scouring and cleaning, a palely glistening edifice, turreted and spired, has emerged with, in front, a new and handsome statue of Robert the Bruce. From the back of his magnificent horse, the noble Bruce holds aloft the Declaration of Arbroath and, as if claiming every honour, every historic and moral victory, on his crown stands *Larus argentatus*, head back, calling, perhaps declaiming every word: 'It is in truth not for glory, nor riches, nor honours that we are fighting but for freedom – for that alone, which no honest man gives up but with life itself.'

The season changes, fast and early. The air one morning is the air of autumn. The difference is inexplicable. It is in the sound in the lanes, the winds stirring the dry whisper of a silver birch, a larch; it's in the berries on the rowans, the plums heavy on their branches, falling over stone walls, dropping neglected in the scent of smoke and peat. It is in our own acknowledgement, too; what had to have been accomplished in summer either has been, or it is too late to be so now. We have to come to terms, to accept, to round up the months of possibility in whatever way we can. Any hopes I had for major works in the garden have been subsumed by rain. No follies, no ha-ha. The laburnum grove won't be planted, the parterre, the vegetable garden to rival Villandry, the reflecting pool. The moss lawn and bog pond will have to do.

August 31st

Inside the house and outside it, moulting is almost over for the year. The crows in Rubislaw Terrace Gardens are at the awkward stage before re-feathering is complete. Chicken too looks on the last day of August like a bird made from bits, from old squares of stuff and tacked-on feathers, dusty grey and fawn all falling adrift, a skinny neck of layered grey fuzz covering fine black quills of unfurled new feather, sharp and intensely black. When held between the hands, she seems to weigh nothing. Is it my imagination that once she was a chunky, solid bird, heavy on the arm or shoulder? (Yes, I think it is. Even now, she feels heavier when standing on my knee.) Birds, I say to myself, are like this: unfixed, movable, imperceptibly altering moment by moment, even in such ways as volume, size and weight, and it is our own perceptions which make us interpret, one way or the other. I don't want to think that she might be lighter through age. In days, she'll be fully occupied releasing her new feathers from their tight white casings, picking and casting the keratin casings to the floor of her house. Even around her

325

eyes, prickles of dark new feathers are needling through. She'll look perfect again, a lovely, feathered rook. I don't know if I'm grateful that birds appear not to show, as humans do, the effects of age. There's still so much to learn. I kneel on the floor and although her eyesight is poor, she comes to me and stands on my wrist, nibbles her beak gently against my fingers.

September 7th

The Dee on the edge of autumn, water rolling like molten glass, dark and green on an afternoon that is mild and grey. Just off the main road, there are paths to the river but we plunge straight through the tall grasses and undergrowth, past soft hollows where roe deer have slept, through banks of tansy and willowherb, cow parsley, meadow grass and Himalayan balsam. The noise of traffic is softened by the trees. This is still a city, a line of suburbs extending beyond us along the far bank of the river.

We perch on rocks in the water for a while then walk downriver. Among the low grass between the rocks we find a dead shrew, or rather I do. This I regard as a triumph. Dave has a raptor's eyes, trained from childhood, seeing more with the naked eye than I do through binoculars. I peer and squint and fiddle with the lenses, and by the time I've focused eye or binocular, half a world of species may have passed before me. This time, I'm the one who finds the creature; a shrew dead in the grass. We stop to examine the tiny form, small and velveteen, one side pierced and bloody. David picks it up by the tail and we walk on with him carrying it, swinging from his fingers. We linger, watch the river on this sunless near-autumn afternoon, the plants and stones and the movement of water.

Back home, I bring the shrew into the house and put him temporarily in the vestibule beside the still drying-out blackbird wings. I lay him on newspaper and turn him over. A common shrew, *Sorex araneus*. The wound in his side may have been made by a creature that, having pierced it, experienced the unpleasant taste given off by the liquid under a shrew's skin. (It's a useful strategy when warding off creatures with a good sense of smell. It is, however, no defence against those who discover the noxious taste having already dealt the mortal blow, which is why so many shrews are delivered by their pets as gifts to cat owners, on the principle, I suppose, of the redistribution of unwanted Christmas presents. The reason that shrews are eaten in quantities by owls is that owls have little sense of smell and, consequently, taste.)

Dead, this shrew is free from the demands that impel his ferocious aggression – the need to defend his territory and to satisfy his voracious eating habits. Shrews eat almost incessantly, a diet of insects and worms, their own weight in a day. I would have to eat 100lbs of food a day to match him, bite for bite, but then I don't have his rampant metabolism, his necessities for warmth and the provision of food for many litters, for constant territorial vigilance and resistance. A shrew's heart beats very fast (in some species, 1,200 beats per minute). Its life, therefore, is short. If they are aggressive, it is because they have to be: a small, one-creature defence force (which shares most of the characteristics of a many-creature attack force). Their destiny is fast-paced and concentrated and now, this creature has met his death just as the season is changing, when he'd have been preparing, physiologically, for winter. Unlike some bird species, the components of whose brain expand in autumn to accommodate the requirements of food-searching and hiding, the shrew shrinks in for winter. Instead of hibernating, he diminishes. He reduces in size, his brain, skull, body, so that there will be less of himself to chill and die.

September 8th

After his night in the vestibule, I bring the shrew into the kitchen and lay him on cartridge paper on the worktop to photograph. Close to, it's clear he has already been dead for a day or so. (I pay particular attention to see if his noisome shrewish propensities are obvious to the naked nose but he smells like nothing more than a decomposing small rodent.) I look at his long, whiskered nose, the glimpse of his small, red, iron-clad teeth and think of his reputation, of Shakespeare's cementing into our culture the negative view of shrews and the long-held superstition that the 'shrew-mouse' is bad luck if seen at the beginning of a journey, the idea which, according to Gilbert White who branded the shrew: 'of so baneful and deleterious a nature that wherever it creeps upon a beast ... the suffering animal ... is threatened with the loss of a limb'. The cure for such suffering was to have the affected part rubbed with the twig of a 'shrew-ash', an ash tree whose trunk had a hole bored in it, into which had been thrust a live shrew. The hole was then plugged up and the shrew left to effect whatever influence it could to ensure that the tree might be used for the aforementioned medicinal purposes.

I think of the ways in which the names of small creatures have been used as terms of endearment or abuse – Shakespeare, destroyer of shrew reputations for centuries, was kinder to mice: 'good, my mouse of virtue', the clown says to Olivia in *Twelfth Night*, while in *Travels With Charley*, during a drunken conversation in a bar in his home town in California, John Steinbeck remonstrates with an old friend who urges him to abandon New York and come home:

'"*Conejo de mi Alma*", I said. "Rabbit of my soul, hear me out".'

There is even Herrick's 'sweet slug-a-bed'; but of this small unfortunate, not a good word is to be found. I photograph him in the limited poses available to us both and after his modelling tasks are completed, take him out to bury him. Small though he is, it is a matter of consideration. Whenever I bury a creature in the garden, I try to remember who is where, to choose a site suitable for age or rank, as if there might be a companionship of bodies – or genus or species – if not of souls. This is a considerable task after so many years of keeping creatures. I remember where the ptarmigan is, a bird from the high Cairngorms in its winter outfit who was picked up (unwisely) by someone who, recognising his poor state of health, decided to bring him from his mountain to me. He's under a white-flowering potentilla, chosen specially. Over the infant jackdaw is a deep-red heuchera. The ground beneath the hellebores harbours our doughty cockatiel and

three pet rats, recently dead, although previous generations are buried
there too, on a now lower level, like catacombs of ancient monks. (It
is an ancestral burial ground. The flat-dwellers in other cities among
the family reverentially return their dead for burial here. This often
involves temporary recourse to the freezer where the incumbent will lie,
suitably shrouded, adjacent to the chips and peas and ice cubes. No
accidents of substitution or consumption have, I believe, ever knowingly taken place.) Apart from plants and the occasional smooth,
rounded stone, which no one would recognise as such, there are no
memorials except for one miniature headstone for Leah's late rat,
Celeste. A present from a friend, it was made from fine local granite by
a craftsman, carved with her initial, polished and gilded and, although
it is an object of sombre reverence, the gesture is not to be encouraged.
I don't want my garden to be macabre, or frankly weird, like that of
Mrs Cornelius Oosthuizen who makes a startling appearance in
Truman Capote's account of his life in Brooklyn, *A House on the
Heights*, pointing out from her window seashells, corals and a cross of
pebbles marking the graves of goldfish, rabbits and canaries, all
described with that edge of unsettling chill at which Capote is master.

I lay the tiny creature in the earth. There's no one else around to take
part although even if there had been, family attendance at these ceremonials has, over the years, alas, become regarded as less than
mandatory. The obsequies are ever brisker. There are even parties who,
like fragile Victorian maidens, find excuses not to attend on the pur-
ported grounds of excessive sorrow while I, as ever, wield the spade. (As
it is, only I know where the bodies are buried.) I've given up providing

ringing eulogies too and instead, raise a brief mental repetition of a verse of Theodore Roethke's poem, 'The Meadow Mouse':

> My thumb of a child that nuzzled in my palm,
> To run under the hawk's wing,
> Under the eye of the great owl watching from the elm tree,
> To live by courtesy of the shrike, the snake, the tom-cat . . .

September 15th

On one of the bright mornings of autumn a friend and I drive out to Forvie to walk. It's a gusting morning of sun and rain and high, chill wind. We walk from Newburgh past a bank of seals, past shags scattered among the flocks of eiders on the water, redshanks, curlews, oystercatchers. We have to close our eyes against the force of the wind. Today we can *see* the wind, feel its quick, fine scouring sandblasting the skin of our faces. (For a day or two after, I'll waken to a dusting of sand grains on my pillow.) A flock of dunlins flies in glorious synchrony through the ice-tinged, blowing air. It was here in 1909 that among the first exercises in modern bird-ringing took place when a student at Aberdeen University, one Arthur Landsborough Thomson, later a distinguished ornithologist, and some friends caught and ringed six lapwings, beginning the study The Aberdeen Scheme, forerunner to the practice that has taught the world so much about the lives of birds. The scheme ended with the beginning of the First World War but other ringing continued later with the founding of the British Trust for Ornithology in 1932. With that was begun what we know of the length of birds' lives, their destinations, their journeys, the nature of their fates.

As we walk back, a pair of swifts dip and rise, wind-carried above us. What are they still doing here? Go, before it's too late, before the frost starts. Temperatures are already edging downwards, fast. It was four degrees centigrade last night. In a few days, snow will fall in the high hills of the Cairngorms. *Fly!*

September 17th

Even now, a couple of days after our walk, I see one or two swifts skimming the low walls and fences over fields on the edge of the city. In the formal garden at Drum Castle, they've gone. The earnest, concentrating faces which wove among and over and round us all summer in their strange and almost ghostly way, over the delphiniums and box hedges, whistling their unearthly messages, are far away now, impelled by that cosmic restlessness that dominates their lives.

September 22nd

Bec's birthday, and the autumn equinox tomorrow, a time of possibly increased geomagnetic storms. Again, I approach the realm of space weather. The nature of my hope is dogged but foredoomed. No activity has been detected, or if it has, no one's telling me.

The branches of the beech tree next door, the ones that stretch a wide roof over the garden so that only hellebores grow under it, are laden with still-green leaves which soon will turn ochre, then copper. On the few days when the sun shines, they light the rooms at the back of the house with a warm, munificent glow but they'll fall like shavings of fine curled metal over the lawn and flowerbeds, from where I will spend weeks chasing them round in a generally futile way with a leaf blower. Because leaves fall serially, are blown by the wind, grow again seasonally and fall again, it's one of the tasks, like dusting or window-cleaning, or indeed any cleaning, that's never actually completed. The tree has grown taller, expanded like a canopy over the garden. (Could the tree be why the phone line is difficult, interrupted; why every now and again the Internet connection disappears into the vapour of nowhere?) One year, I went away for a few days in mid-October. The beech tree was still in leaf when I went but when I came back it was leafless, the garden evenly ankle-deep in brilliant gold, as if some deranged leaf-showering fairy had suffered a severe episode while I was away.

September 28th

The first evening of Rosh Hashanah, the New Year. This, contrarily, is a beginning, the festival when we dip apples in honey in the hope of invoking sweetness for the coming year. I do it in the generally religiously unobservant way that notes ritual and time, that bows to the year that winds inside my head, only one single strand of the complex timelines of my life, one of the memories of outgrown lives that we all still carry with us, every pattern of the past, school terms and academic years, years of child-rearing and of work, the festivities and dates and occasions by which we mark our lives. The words of the Rosh Hashanah prayer, the *shehecheyanu*, are beautiful and feel prescient: 'Blessed art thou oh Lord our God, King of the Universe, who hast kept us in life, and hast preserved us, and enabled us to reach this season . . .'

October 2nd

The first days of October. Gushing rain and high wind. As I work, *Passer domesticus* sing in chorus from the viburnum with a vigour that shames me. In the garden, everything's closing, turning, falling. Angles alter, stems tip and bend, draw down towards the earth, begin to wither. The leaves fall from the *Hydrangea petiolaris*, or hang against the branches, a pale bleached-out yellow. I draw a rake lightly across the surface and the leaves drop limply from the stems.

October 11th

It grows colder. For all the change in the sense of air, there's no one moment when autumn elides coolly into winter. Even the change in clocks will be late this year. The darkness will have already caught up with time. We think that, as a species, *Homo sapiens* has moved beyond the immediacy of seasons. We alter clocks, play with time and light, forgetting that it is a mere adjustment, that both are within the workings of the larger world, both fixed and finite, ultimately unalterable.

Of all seasons, autumn is the one that reminds me, if ever I need reminding, that even as a human being, supposedly freed from some of the burdens of nearness to 'nature', I'm allowed no dispensations, that I'm no different from any other species, tied as we all are, inescapably, to the fabric of the universe. Towards autumn, we all, human or not, begin to make preparation for winter. Here, we settle into the cold stone of our houses, begin the often unceasing war against the material of which our only earthly shelters are made; here, granite which shares heat-keeping properties with wet denim and flooded waders. Walls are thick, windows thin. Across the North Sea, the homes of our nearest continental neighbours in Norway, Denmark and Sweden are built from wood and are enviably warm.

In preparation for winter we have, according to a nature guide for Michigan teachers, three alternatives: 'migration, dormancy, or toughing it out'. (That this refers to birds and animals in the north seems to me to be irrelevant. Useful advice is useful advice.)

Outside, there is collecting and storing and within us all, the changes that ready us for cold. By now, wild birds too have finished moulting and have regrown their winter feathers. The migrators pass overhead, seen or unseen but heard like the geese behind the cloud layer, heading north.

Not only migrating birds; we're all on the move. In every way, we are. We change, all of us, human, bird, animal; externally, internally, in concert, in rhythm, knowingly or unknowingly with our surroundings, our moods and inclinations. Our clothes, fur, feathers, metabolism change and if we belong to species that live outside, more urgently, the sources and supply of our food change, our potential security, our homes, our familiar landscapes. If we don't have the urge to stay, to hide, to sleep, close in on ourselves, we may experience the overpowering, irresistible urge to go, to fly, to embark upon our annual, or biennial escape, for warmth or food, or the fleeting desire for many of us if we live in the north, not to be here. We grow or we shrink. We store up for the winter, newly feathered or newly fattened. In preparation for migration, birds' metabolism changes. They have to eat up to 30 per cent more than usual. Their fat production and storage increases as they lay in the fuel for their often non-stop journeys.

Some of our brains, or parts thereof, may grow. The hippocampus (the memory area in the brains of both humans and birds) of 'caching' birds expands in autumn. 'Caching' birds hide and store food all year round but with the approach of winter, increase their storing. Corvids, the main avian 'cachers', will have many cache sites, and being habitual pilferers who develop elaborate manoeuvres to outsmart other corvid thieves require considerable memory to guard and remember their stores. (Seasonal brain plasticity in birds is demonstrated in spring too, when the parts of the brain involved with song-learning and control expand.)

336

The brains of others shrink. Those of the shrew and other small mammals: grey squirrels, ferrets, some species of voles, hamsters, dormice, mice and bats. In rodents, the brain comprises two to three per cent of the body mass but requires the energy expenditure of 10 per cent and so any winter reduction of energy requirements increases the chances of survival. Shrews born in early summer experience a brain-size reduction of over a quarter before winter – their basal ganglia reduce by 29.8 per cent, their neocortex by 27.5 per cent. (You may relinquish your more profound thoughts. Who needs them in winter, anyway? Hegel, anyone?) The size of some birds' digestive tracts expands in winter, contracts in spring. (The brains of migratory birds have been found to be smaller than those of non-migrators, probably because migration requires a reduction in the high-energy costs of having a larger brain.)

In autumn, some animals and even birds enter states of hibernation or torpor – a less prolonged state than hibernation – when body temperature and metabolic rates are reduced and respiratory rates fall. Swifts in a state of torpor take only eight breaths a minute.

We're not as different as we think we are – if we don't actually enter a state of torpor, the tug is there from desire, negation, reluctance to face the winter. There's nothing spiritual, nothing mystical, unexplained, supranormal in the knowledge: we are one with everything else on this vast earth that sleeps and wakens as the course of the seasons makes its progress. We're formed from the same strange amalgam of cells, tied in ways we don't quite understand to the molten core of the earth, to the moon and tides and light; the rhythms, the circadian, the

24-hour cycle in biochemical, physiological and behavioural processes in living beings, plants, animals, birds, the infradian, that once-in-28-days cycle of menstruation and tides, the ultradian, the one by which our heart beats.

Where is volition? We're as subject to the forces of those stern Teutonic poetics, *Zugunruhe, Zeitgeber* as anything else – *Zugunruhe*, that cosmic restlessness, the urge to migrate, and *Zeitgeber*, the external stimulus, most often light, that orders our internal clocks.

During winter, above certain latitudes, we all may be sad. Some are not only sad but afflicted by a named phenomenon with a suitable acronym: SAD, or Seasonal Affective Disorder, that sense of cold and light-denied despair, that pervasive mental ice-cloud which falls with extending winter darkness over the spirits of the inhabitants of the northern areas of the earth. There are many names for it. It's *kaamos* in Finland (a word that also means 'polar night'); in Dutch, it's *somber*, in German *finsternis*, in Italian *malinconia*. The Spanish suffer from *tristeza*, the Swedish from *skugga*, a word meaning 'shadow' or from *dysterhet*, melancholy, or *somber*. Latvians and Lithuanians suffer *rudens depresija*, autumn depression. The Western Isles and west coast of Scotland are prey to the 'Highland gloom' (a darker and more pervasive gloom than the usual workaday Scottish variety), and in Orkney and Shetland, the northernmost islands, to a psychiatric ailment once known as *Morbus Orcadiensis*. Such poetry of grim! The term *depressio hiemalis* is useful because it covers the lot – winter depression. In Barry Lopez's *Arctic Dreams*, he describes an Inuit form of the phenomenon, *perlerorneq*, 'the weight of life', a more serious and psychotic response to perpetual darkness, which causes people to run half naked into the cold, screaming, or to tear at their clothing in despair. In whatever language, the words carry the same import: crepuscular, dark, melancholic, obscure – the obscurity, or obscuring of oneself, the overshadowing of optimism or a sense of future, an ontological disappearance into the vast, incalculable maw of winter darkness, into the chill that is the negation of life, that echo and shadow of the darkness of death.

(There is a related malady, one not tied perhaps as strictly to latitude but more, perhaps, to culture. It might be called *petit* SAD to winter's *grand* SAD. SAM – Sunday Afternoon Malady – the thought of Monday and another week. I think of the enforced jollity of Sunday-afternoon radio when I was a child, a jollity we always knew to be legerdemain, a bluff to pretend to us that, eschatologically speaking, all would end well when, even as small children, we knew very well that it never does.)

October 15th

A friend has sent me a poem he has translated from the Chinese. It is the great T'ang poet Du Fu's 'Autumn Meditation' and is translated into Scots. He calls it, 'Hairst-Time Quicknin'. It is beautiful. I smell autumn from it but an autumn from the past, language as a reliquary, as the censer of which Du Fu writes, sending out the slow savours of time long gone. Du Fu and his eternal sadness:

Wind blows clouds across the pass casting darkness to the
earth
Sprays of twice-blossoming chrysanthemum recall tears
from other days . . .

Du Fu, I want to say to him, your sadness is only in part to do with
your perceived failures, with your political instabilities, the horrors of
the An Lushan Rebellion, the hardships of your life, your sense of duty
and compassion. I've made a diagnosis. In spite of your southern latitude, you've got SAD, I want to tell him, possibly comfortingly, over
our lost centuries; it's all, or some of it, to do with the season and the
light.

On the website of a company selling equipment to help overcome
the effects of SAD – the special lights, light boxes and clocks which
bring reluctant wakeners, those who might otherwise wish to remain
in the perpetual hibernation of winter sleep, gradually into the glowing, healthy dawn of a SAD-free day – one of the symptoms of the
affliction is described as: 'More than normal sadness', and as I read this,
I bow to the wisdom of the philosopher-merchant, the cool-eyed savant
who wrote it, the one who, philosophically at least, has diagnosed us
all and come with calm and sensible acceptance to a profound understanding that this is just the way life is. 'Normal sadness', one of the
underpinnings of our lives, just one of the states of our being, the
bedrock from which, with difficulty and fortitude, in the presence eternal of boredom and joy, death and fear and sin, we rise and continue,
womanfully, manfully, to live.

Inside the house, the birds are prepared for a winter they will not face, or not in the way their wild counterparts will. If there's any more caching than usual, I don't notice. (There has to be an instinctive motive for caching, so perhaps with my corvids it's a recreational one. Neither Chicken nor Ziki need to store food but both do, enthusiastically. It takes up a good deal of time. It might – if one had very few other calls on one's time and enjoyed being involved in games of hiding, stealing and retrieving – be fun.) They are properly feathered and with central heating, they will be considerably warmer than anyone else.

The leaves on the beech have begun to fall in great storms of gold. The tree will be leafless in days. Wood pigeons and collared doves will distribute themselves among the empty branches, perching ever higher during the day in the dappled patches of sun-strike. They'll be there outside the windows of every floor as I look out.

The earth turns. My heart sinks momentarily with the prospect of darkness, with late afternoon. I don't know why because I prefer winter and welcome the clocks changing, the feeling of regaining that snatched and stolen time. My hour is precious and given back to me late this year, at the very end of the month. By the time it happens, it is no longer a herald of winter because winter began weeks ago. The clocks will change and I'll savour the atavistic pleasure of moving very slightly back into the past, of regaining time, being given the gift of an hour, and darkness.

October 25th

On a day of fierce, bitter winds, I see a wood mouse, *Apodemus sylvaticus*, scaling the clematis outside the study window. I thought I had a hint of his arrival a few days ago when I found some birdseed husks in a neat heap on the step. Years ago, we found a wood mouse feeding in a jar of bird food. We admired him through his wall of glass then tracked the speed with which he returned from his place of release at the bottom of the garden to the house. Identifiable by his truncated tail – wood mice shed their tails when caught by them – this particular one, half-tailed and vigorous, must have sprinted up the garden like a rodent Olympic sprinter because he was seen within minutes of release, back crossing the study floor.

Craig and Dod, on finishing their work, took away the last of the poison from under the house. This particular rodent, coming in for shelter, will be safe. Unlike house mice, they leave little trace of themselves. I won't do anything to discourage their presence so modest are they in their demands and their effects. All winter now and into spring, I'll find a single Rice Krispie taken from the floor beside Chicken's dish, or a tiny piece of bread hidden behind a cushion or left in a corner. Once or twice as I work, startled by the sound of a large and vigorous animal in the room, I will take a torch and shine it into the tunnel between the wall and the cupboard in my study and see the miscreant – a tiny, big-eared mouse pressed against the skirting board, huge-eyed in the torchlight. Now we will spend the winter together, sharing space and air.

November 2nd

Three days into winter with the shunting of the light. Afternoons seem briefer, hours altered to fit the quickening of night. This afternoon, I walk in a circuit past the harbour and the Northern Isles ferry across Victoria Bridge, through Torry to look out over the sea and back towards the deep, smudged triangle of stone and light, charcoal and flake white, the city in a pearly haze of unfallen rain. Dolphins and porpoises, four of each, swim beyond the harbour channels, in and out of the waves in momentary darts of sunlight.

What is it that makes the year different when the clocks change, slants us northwards, closes the world down? It feels like an acknowledgement of place, the brave embrace of inevitable winter. I turn from the red-throated divers and guillemots, the mallards and eider by the breakwater to walk back. Darkness an hour earlier, the shop lights and lit windows ahead newly welcome, a sign of comfort in human presence. The streets here are all reminders of haven and darkness, of refuge from a wild, cold sea. I walk up Market Street past fishermen's rests and missions to seafarers, the name above a door, 'Apostolatus Maris', past bars and pubs, cheap hotels and rooms rented by the hour, an enigmatic shopfront bearing the name 'The Ecstasy Annexe', reminding me of what impels us all, the lives of birds and beasts and man.

Almost a year since the beginning of the snow. A string of Christmas lights frames a dusty window, brightly hopeful. As I walk, a cloud of gulls rises from its frantic pickings among the fish-house bins and spreads into the nearly evening air like a sudden storm of early snow.

Acknowledgements

I am grateful to many people who helped me in the writing of this book.

To Joyce Gunn Cairns, a wonderful artist, and friend, I give my most profound thanks. Her wisdom, erudition, insight and humour have cheered and encouraged me at every turn, while her exquisite drawings of crows, spiders, mice and rats (and people) have demonstrated to me at every moment the finest, most insightful ways there are of understanding and interpreting the beauty of the natural world.

I am grateful to everyone at the Aberdeen University's Centre for Environmental Sustainability – in particular, Steve Redpath, Andrew Whitehouse and Mark Reed. Thank you too to Huw Warren, Matthew Dalziel and Louise Scullion, to Sera Irvine and Helen Denerley, my fellow-residents at ACES for their company, their brilliant conversation and ideas. To Helen I owe a most special kind of gratitude for her unflagging support and constant encouragement, her indomitable spirit and the inspiration of her own art. (I think too of Molly, of the peesies, curlews and long-eared owls of Clashnettie, and of the presence and memory of Peter Welch which will be there always in the garden and the house and in the quiet, peaceful air above the Deskry.)

ACKNOWLEDGEMENTS

My thanks to Aberdeen Harbour Master Captain Roy Shaw and Dianne Insch, who were generous with their knowledge and their time, as were Hugh Black and Sandy Whyte, owners of Rubislaw Quarry.

My agent Jenny Brown has been wonderful, as ever, and to her and to everyone at Granta, I give many and most sincere thanks. To Bella Lacey, Sara Holloway, Max Porter, Pru Rowlandson, Christine Lo and Brigid Macleod, I owe huge gratitude, as I do to Jenny Page and Benjamin Buchan. Mónica Naranjo Uribe's delightful cover, endpapers and illustrations have made this book more beautiful than I could have hoped for.

The insight, kindness and company of my cousins Roger Alexander and Hazel Woolfson have been a constant source of comfort and cheer. Ross Whyte, Neil Skinner, Lee Carr, Colin Glasser and Gill Owen have all been stalwart in their help and support. For 'Hairst-Time Quicknin' (and the amazing depth of knowledge behind it) I give thanks to Brian Holton.

And at last, much gratitude is owed to Bernard Krichefski, after all these years.

To Dave, Gillian, to Bec and Leah, Han and Ian, my love and thanks always.

Select Bibliography

Allen, Barbara, *Pigeon* (London, 2009).

Anderson, Lorraine and Edwards, T.S., *At Home on This Earth: Two Centuries of U.S. Women's Nature Writing* (Hanover, New Hampshire, 2002).

Balcombe, Jonathan, *Second Nature: The Inner Lives of Animals* (New York, 2010).

Barker, Elspeth, *O Caledonia* (London, 1992).

Bekoff, Marc, *Minding Animals: Awareness, Emotions and Heart* (Oxford, 2002).

Bekoff, Marc and Pierce, Jessica, *Wild Justice* (Chicago, 2009).

Beston, Henry, *The Outermost House* (New York, 1988).

Carroll, Sean B., *Endless Forms Most Beautiful: The New Science of Evo-Devo and the Making of the Animal Kingdom* (London 2005).

Cronon, William (ed.), 'The Trouble with Wilderness; or, Getting Back to the Wrong Nature,' in *Uncommon Ground: Rethinking the Human Place in Nature* (New York, 1995).

Davis, Mark A., *Invasion Biology* (Oxford, 2009).

Darwin, Charles, *The Formation of Vegetable Mould Through the Action of Worms with Observations on Their Habits* (London 1881).

De Bary, William Theodore, *Sources of Chinese Tradition*, vol. 1 (New York, 2001).

Dillard, Annie, Pilgrim at Tinker Creek (London, 1974).

Ellis, R.A., *Spiderland* (London, 1912).

Foster, Russell and Kreitzman, Leon, *Seasons of Life* (London 2009).

Gianquitto, Tina, 'Good Observers of Nature': *American Women and the Scientific Study of the Natural World* (Athens, Georgia, 2007).

Gould, Stephen Jay, *I Have Landed* (London, 2002).

Halle, Louis J., *The Appreciation of Birds* (Baltimore, 1989).

Hillyard, Paul, *The Private Lives of Spiders* (London, 2011).

Hull, Robin, *Scottish Mammals* (Edinburgh, 2007).

Hutchinson, G.E., 'Homage to Santa Rosalia or Why Are There So Many Kinds of Animals?', *The American Naturalist*, vol. XCIII, no. 870 (May-June, 1959).

Lockwood, Jeffrey, 'The Nature of Violence', *Orion Magazine* (January-February, 2006).

Lorenz, Konrad, *On Aggression* (London, 1966).

Lorenz, Konrad, *King Solomon's Ring* (London, 1953).

Lovegrove, Roger, *Silent Fields: The Long Decline of a Nation's Wildlife* (Oxford, 2007).

Louv, Richard, *The Nature Principle* (New York, 2011).

McKibben, Bill, *Eaarth* (New York, 2010).

Maclean, Norman (ed.), *Silent Summer: The State of Wildlife in Britain and Ireland* (Cambridge, 2010).

Marren, Peter, *A Natural History of Aberdeen* (Finzean, Aberdeenshire, 1982).

Mayr, Ernst, *Systematics and the Origin of Species* (Harvard, 1942).

Preston-Mafham, Ken and Rod, *The Natural History of Spiders* (Marlborough, Wiltshire, 1996).

Sax, Boria, *Animals in the Third Reich* (London, 2000).

Sloane, Barney, *The Black Death in London* (Stroud, 2011).

Snyder, Gary, *Back on the Fire* (Berkeley, 2007).

Snyder, Gary, *A Place in Space* (Berkeley, 1995).

Steingraber, Sandra, 'The Fall of a Sparrow', *Orion Magazine* (January-February, 2008).

Tinbergen, Niko, *The Herring Gull's World* (London, 1953).

Tudge, Colin, *Consider the Birds: Who They Are and What They Do* (London, 2008).

Tallmadge, John, 'Resistance to Urban Nature', *Michigan Quarterly Review*, vol. XL, no.1 (Winter 2001).

Watts, Alan (with Al Chung-Liang Huang), *Tao: The Watercourse Way* (London, 1975).

White, Gilbert, *The Natural History of Selborne* (London, 1977).

Wilson, E.O., *Biophilia* (Harvard, 1984).

Yu-lan Fung and Bodde, Derk, *A History of Chinese Philosophy* (Princeton, 1983).

Text Credits

The author and the publisher have made every effort to trace copyright holders. Please contact the publisher if you are aware of any omissions.

Lines from *Venice Revealed* copyright © Paolo Barbera. Reprinted with the kind permission of Souvenir Press and Laura Morris.

Lines from *Wild Justice: The Moral Lives of Animals* by Mark Bekoff and Jessica Pierce copyright © University of Chicago Press 2009.

Lines from *The Outermost House: A Year of Life on the Great Beach of Cape Cod* by Henry Beston copyright © Elizabeth C. Beston 1977.

Lines from *Pilgrim at Tinker Creek* reprinted by the permission of Russell & Volkening as agents for the author's estate. Copyright © 1974 Annie Dillard.

Lines from *At Home of this Earth: Two Centuries of U.S. Women's Nature Writing* by Thomas S Edwards and Lorraine Anderson copyright © University Press of New England, Lebanon, NH. Reprinted with permission quotes from page 127.

Lines from *A Parliament of Birds* copyright © John Heath-Stubbs. Reprinted with the permission of David Higham Associates.

Lines from *The Lost Notebooks of Loren Eiseley* copyright © Kenneth Heuer and the Estate of Mabel L. Eiseley. Reprinted with the kind permission of the publisher, Little, Brown and Company.

Lines from 'Homage to Santa Rosalia or Why Are There So Many Kinds of Animals?' copyright © G. E. Hutchinson. This essay first appeared in *The American Naturalist*, Vol. XCIII, No.870 (May-June, 1959). Published by University of Chicago Press.

Lines from *What the Stones Remember* by Patrick Lane © 2004, 2005 by Patrick Lane. Reprinted by arrangement with Shambhala Publications Inc., Boston, MA and McClelland & Stewart.

Lines from 'The Nature of Violence' copyright © Jeffrey Lockwood. This essay originally appeared in *Orion* magazine, January / February 2006.

Lines from *King Solomon's Ring*, p 146, by Konrad Lorentz copyright © Routledge 2002. Reprinted with the permission of Routledge.

Index

Aberdeen
  19th-century plant life, 98–9
  appearance, 18, 19–20
  atmosphere, 20–2
  beach nearby, 44–8
  The Chanonry, 177
  city-centre under-streets, 115
  claims to fame, 25
  coldness, 94, 95
  Corbie Heugh, 284
  the Green, 115
  harbour and quays, 52–3, 221–3
  High Street, 175–6
  history, 96–7
  layout, 95–6
  Marischal College, 324
  Marischal Street, 52–3
  Market Street, 343
  Old Aberdeen, 175–6
  Robert the Bruce statue, 324
  Rubislaw Den, 33, 98–102
  Rubislaw Quarry, 237–44
  Seaton Park, 178–80
  situation, 18
  Union Terrace Gardens, 283–4
Aberdeen University, 59
  Zoology Museum, 277–8
African Americans: relationship with countryside
    and wilderness, 170–1
Akan peoples, 123
Aldrovandi, Ulisse, 89, 90
Allen, Barbara, 178
Allen, Woody, 181
Altenburg, 249
Amsterdam, 272
Andrew (bird walk leader), 175, 177, 179,
    180
Angell, Tony, 284–5
animals
  capacity for violence, 113–17
  hierarchy of creation, 67–8
  human infliction of pain on other, 212–16
  human relationship with other, 199–201
ants, 124, 202–3, 215
Aquinas, Thomas, 67

arachnophobia, 124–6, 128, 137
Aristotle, 67, 89
Arrhenius, Svante August, 231
Atlantis space shuttle programme, 264
Audubon, John James, 177, 276, 277–8
Augustine, St, 67
aurora borealis (Northern Lights), 23–7, 264,
    265–6, 287
'Autumn Meditation' (Du Fu), 339–40

bacteria, 106–7
Banksy, 69
Barbaro, Paolo, 119
Bardie (cockatiel), 28, 42, 195–8
Barker, Elspeth, 252
bats, 263
Bayern, Auguste von, 254
Bec (author's daughter)
  and birds, 255
  birth, 286
  birthday, 110
  houses lived in, 22
  pets, 78, 195–8
  visits to author, 41
Bede, the Venerable, 282
Bedford, Sybille, 93
beech trees, 333
bees, 298, 317
beetles, 262
Bekoff, Marc, 75, 76
Belgrade, 272
Beltrá, Daniel, 159
Berendt, John, 190–1
Berlin, 272
Bernstein, Dr Aaron, 107
Beston, Henry, 45, 105, 159–60
Bible, 202, 235
biodiversity, 105–9
  and latitude, 110–12
biophilia, 201–2
birds
  ageing and appearance, 195–6
  bathing in water and sun, 259–60
  bird walk in Aberdeen, 175–80
  challenges to urban, 146–9

cognitive skills, 144–6, 148, 251
'dialects' in calls, 147
diseases and hygiene, 54–5, 190–1
extinctions, 275–82
fear of, 125, 128
fidelity to roosting sites, 283–5
human relationship with wild, 199–200
longevity, 196
migratory, 313, 336, 337
monogamy, 185–6, 252–3
moulting, 325–6
navigation skills, 119, 120
and oil spills, 157–60
protecting and rescuing, 3–9, 44, 40, 154–5,
    156–71, 203–5, 245–56
ringing schemes, 331–2
seasonal brain plasticity, 328, 336
smell of, 248
suborders, 114
varieties on beach near Aberdeen, 46
see also individual birds by name; nests and
    nesting
Black, Hugh, 243–4
Black Death, 68–9
blackbirds, 203–4, 255, 271
blue tits, 112–13
Bouchet, Philippe, 208
Bourgeois, Louise, 132–3
brain plasticity, seasonal, 328,
    336–7
Brian (Magic Roundabout), 209
'Burbank with a Baedeker' (Eliot), 70

Capote, Truman, 330
cars, 280–1
Carson, Rachel, 92
cats, 225, 273
Catullus, 281–2
Celeste (rat), 330
Chadd, Richard, 136
Champlain, Samuel de, 93
chanterelles, 262
Charlotte's Web (White), 130–1
Chase, Dr Ronald, 211
Chicken (rook)
    age, 28
    appearance, 201
    and family jackdaw, 250
    feeding, 249

habits and behaviour, 38, 39–40, 42, 61, 341
    moulting, 325–6
    nesting, 143–4, 157, 164, 195
chickens, 145
China
    attitude to rats, 69
    pest control campaign, 278–9
Chivian, Dr Eric, 107, 236
Christmas Rose, 89
Church, Russell, 75
cities
    challeμnges posed to birds, 146–9
    as heat islands, 93–4
    vs countryside, 149–54
civil rights movement, 170
climate change, 92, 135–6, 230–5
clocks, changing the, 110, 341, 343
cockatiels see Bardie
Conniff, Richard, 106–7
conservation
    Aberdeen conference, 322–4
    derisory terms for conservationists, 282–3
    extinctions, 275–83
    selectivity of human approach, 200–1
Cordyline australis, 57–8
cormorants, 222, 321
corvids see crows; jackdaws; ravens
countryside
    differing attitudes to, 164–73
    vs cities, 149–54
Craig (rat control man), 64, 66, 70, 80, 342
creation, hierarchy of, 67–8
Creshkoff, Rebekah, 146–7
Cronon, William, 150
crows
    ability to recognise people, 148, 251
    appearance, 128
    fledglings, 155
    food caching, 336
    human dislike of, 64, 67, 128
    roosting places, 284–5
    talking, 286–7
    see also Ziki
Crutzen, Paul, 232–3
curlew, 48

Darling, Frank Fraser, 302
Darwin, Charles, 108, 130, 183–4, 214–17,
    298

INDEX

David (author's husband)
  and birds, 245, 246, 248, 286
  carvings by, 85
  as doctor, 286
  eyesight, 326
  houses lived in, 22
  mushroom-collecting, 262
  and rabbits, 283
  walk with author, 326–7
Davis, Mark A., 294–5
Deakin, Roger, 229–30, 306
Dee, David, 173
Dee river, 33, 326
Deepwater Horizon drilling rig, 157–8
Denburn, 33, 95, 242
Descartes, René, 67
Dillard, Annie, 186
diseases, zoonotic *see* zoonotic diseases
Dod (rat control man), 64, 66, 70, 80, 342
Doerr, Anthony, 233–4
dolphins, 223, 311, 343
Donne, John, 40
Douglas, Mary, 71–2
doves, 182–92
  in author's doo'cot, 42, 86–7, 188–9, 225–6
  bathing in sun, 259–60
  collared, 61, 187–8
  predators, 225–6
  in winter, 341
Dresden, 272
Drum Castle, 160, 332
Du Fu, 339–40
dunes, 46
dunlins, 331
dust and dusting, 121

earth
  as seen from moon, 265
  as seen from plane, 264–5
eclipses
  lunar, 37–40
  solar, 43
ecology, 105–9
Edinburgh
  Bruntsfield Links, 245
  Mary King's Close, 73
  Middle Meadow Walk, 286
  Royal Infirmary, 285–6
  sparrow numbers, 272

Eiseley, Loren, vii
El Escorial, 324
Eliot, T. S., 70
Ellis, R. A., 131, 132
Elmo (rat), 77–8
Elton, Charles, 294
Emery, Professor Nathan, 145, 254
Evans, Karl L., 149
Eversham, Brian, 136
evolution
  and cooperation, 75–6
  and perfection, 304–5
extinctions, 275–83

fieldfares, 279
Fitter, Alastair, 92
Fitter, Richard, 91–2
forsythia, 87–8
Forvie, 48
Foster, Russell, 93
Fourier, Joseph, 231
foxes, 226–8
Frisch, Karl von, 317
frogs, 260–1
fulmars, 321

Galileo Galilei, 25–6
gannets, 321
gardens
  development of author's, 83–8
  function, 279–80
Gaston, Kevin J., 149
geese, 118, 120
genomics, 74
Gessner, Konrad, 89, 90
Gianquitto, Tina, 138
Giant African land snails, 208
Gibbon, Lewis Grassic, 94
Glasgow, 19, 119, 151
global warming *see* climate change
Goodall, Jane, 252
Gould, Stephen Jay, 304–5, 316
graffiti artists, 69
granite, 239–41, 244, 254, 335
Great Chain of Being, 67–8
great tits
  ability to recognise reflections, 143, 144,
    145–6, 157, 160
  dialects in calls, 147

fledglings and death, 198, 203
nests, 195
greenfinches, 155–6
grief, animals' capacity to feel, 186–7, 252
gryllacridids, 116–17
guillemots, 109
gulls, 314–24
    appearance, 33, 52
    black-headed, 243
    calls, 17, 319–20
    classification, 315–17
    distinguishing species, 322
    eating habits, 323
    habits and behaviour, 317–19
    mating and breeding, 318–19
    threatened species, 270
    urbanisation, 314–15
    varieties round Aberdeen, 222

haar, 118
habitats
    animals' devotion to, 283–6
    destruction of, 279–80
hairiness, 125–6
Haldane, J. B. S., 109
Halle, Louis J., 229
hamsters, 73
Han (author's daughter)
    birthday, 110
    houses lived in, 22
    pets, 77–8
    and rabbits, 283
    and spiders, 137
    and swallows, 248
    visits to author, 41
Hatto, Bishop, 69
havdalah, 266–7
hawfinches, 100–1
hawks, 204, 225; see also sparrowhawks
heat islands, 93–4
Helen D (author's friend), 44–8
hellebores, 89
Hemiptera, 262
herons, 261
Herrick, Robert, 329
Herzen, Alexander, 170
hibernation, 337
Hillyard, Paul, 125, 136
Himalayan balsam, 298

histoplasmosis, 54–5
house and land purchase, Scottish system, 239
house martins, 228–9
Huang, Al Chung-Liang, 31
Hull, Robin, 302–3
humans
    effect of seasonal changes, 337–40
    pain infliction on other animals, 212–16
    place in natural order, 67–8
    relationship with other animals, 199–201
Hutchinson, George Evelyn, 107–9, 303
Hydrangea petiolaris, 87, 144, 334

Ian (author's family), 41
icicles, 30
Imbolc, 60
India: sparrow numbers, 272
insects, 106, 109, 116–17
International Union for the Conservation of
    Nature (IUCN), 275–6
invasion biology, 292–307
Iran, 265
Israel
    frogs in, 261
    Jerusalem, 202
    kibbutzes, 171–2
    Tel Meggido, 97
IUCN see International Union for the
    Conservation of Nature
ivy, 88

jackdaws, 245–56
    appearance, 128, 154
    cognitive abilities, 251
    eyes and sight, 254–5
    food caching, 336
    habits, 39, 251–2
    hatching, 253–4
    human dislike of, 154–5
    mating, 252–3
    nests, 247
    rescuing fledglings, 245–56
Jainism, 139–40, 236–7
Jerusalem, 202
Jews and Jewishness
    attitude to countryside, 164–70, 171–3
    attitude to sport, 173–4
    havdalah, 266–7
    migration, 167–8

INDEX

Jews and Jewishness – *continued*
    Passover, 163–4, 165, 174
    Rosh Hashanah, 333–4
Juvenal, 173–4

Kennedy, John, 119–20
kibbutzes, 171–2
Kielder Forest, 304
kittiwakes, 321
Kreitzman, Leon, 93

LaFage, Dr Jeff, 117
Lane, Patrick, 138, 186
Lao Tzu, 235
latitude, and biodiversity, 110–12
Le Rat, Blek, 69
leaf clearing, 333
Leah (author's granddaughter), 41, 189–90, 246, 330
Lenten Rose, 89
Linnaeus, Carl, 89, 90
Llull, Ramon, 67
Locke, John, 216
Lockwood, Jeffrey A., 116–17
locusts, 124
Lopez, Barry, 338–9
Lorenz, Konrad, 249–50, 251, 252, 317
Louv, Richard, 281
Lovegrove, Roger, 272, 302
Lundy Island, 272
Lyell, Charles, 183–4

MacCaig, Norman, 153, 270–1
MacGillivray, William, 176–7, 263, 277, 278
McKibben, Bill, 276
Maclean, Norman, 92
MacLean, Sorley, 153
magpies, 64, 113–15, 187, 200, 204, 273–4
Maklakov, Alexei, 147–8
Malebranche, Nicolas, 68, 185–6
Manx shearwaters, 321
Mao Tse-tung, 278
Marquis, Don, 130
Marren, Peter, 179
Marsham, Robert, 90–1
Marx, Leo, 149
Marzluff, John, 284–5
Matisse, Henri, 183
Matthews, Anne, 146

Max (starling), 55
mayflies, 262
Mayr, Ernst, 316
'The Meadow Mouse' (Roethke), 331
mice, 329, 331, 337, 342
migration, 167–70
mink, 225–6
mists, 118–19
mockingbirds, 148
molluscs, 206–18
    cognitive abilities and capacity to feel pain, 212–16
    mating, 211–12
    purpose, 209–10, 217–18
moon: eclipses, 37–40
Morgan, Edwin, 56–7
Morren, Charles, 89
murres, 105, 109
mushrooms, 262–3

natural events, timing of *see* phenology
nature, human relationship with, 199–202
navigation
    aerial, 119–20, 184
    slugs, 210–11
Nazis, 70
nests and nesting
    encouraging, 112–13, 198–9
    great tits, 195
    jackdaws, 247
    pet rooks, 143–4, 157, 164, 195
    swallows, swifts and house martins, 229–30
New York, 146–7
Northern Lights *see* aurora borealis
'The Northern Lights of Old Aberdeen' (song), 26–7
Nowak, Martin, 76

O Caledonia (Barker), 252
Odum, Eugene, 265
oil spills, 157–60
Oliver, Mary, 278
Oosthuizen, Mrs Cornelius, 330
otters, 186
owls, 327
oystercatchers, 58–60, 160, 320

Passover, 163–4, 165, 174
pastoral idealism, 149–54

pelicans, 159
pets
    problems with small, 72–3
    rats as, 70–2, 73–4, 77–80
phenology, 89–93
Picasso, Pablo, 183
Pierce, Jessica, 75
pigeons, 180–92
    capacity for grief, 186–7
    control, 190–1
    extinction of passenger pigeons, 276–7
    hatred of, 177–8, 180–1
    homing instinct, 184–5
    monogamy, 185–6
    origins, 182
    rescuing fledglings, 3–9, 33, 40
    roles, 182–3
    roosting, 52
    in Venice, 189–92
    in winter, 341
    wood pigeons, 187–8
'Pity the Poor Spider' (Marquis), 130
plague, 68–9
Plato, 67
Pliny the Elder, 89, 298
pond life, 259, 260–2
porpoises, 311, 343
Porter, Gene Stratton, 276–7
Prague
    Old Town bridge, 135
    sparrow numbers, 272
Preston-Mafham, Ken and Rod, 132

rabbits, 124, 283, 329
Rajasthan, 69, 265
rats, 62–80
    as experimental subjects, 74–7
    fecundity, 66
    origins, 65
    as pests, 62–4, 66–9, 80
    as pets, 70–2, 73–4, 77–80
    popular view, 64, 69
    population size, 66
ravens, 126, 128
    food caching, 336
reflections, ability to recognise, 143, 144–6
Reid, Alastair, 50
religion
    Scottish attitude, 49–51

    see also Jainism; Jews and Jewishness
rhododendrons, 299
Robert the Bruce, 324
robins, 28, 200, 201
Rocroi, Jean-Pierre, 208
rodents, brains, 337; see also mice; rats
Roethke, Theodore, 331
Rohde, Professor Klaus, 106
Ronson, Jon, 174
rooks, 29–30, 259, 283–4; see also Chicken
Rosalia, Santa, 107–9
roses, 84, 88, 297
Rosh Hashanah, 333–4
'Rowan Tree' (song), 85
Rupert (rat), 70–1
Rutte, Claudia, 75–6

SAD see Seasonal Affective Disorder
sanderlings, 47
Sax, Boria, 70
Scotland
    attitude to landscape and countryside, 151–4
    attitude to weather and religion, 49–52
    climate, 16
    house and land purchase system, 239
    Jewish immigration, 167
    pleasure and morality, 154
    urbanisation, 151
'Scotland' (Reid), 50
seabirds, 222, 321; see also gulls
Seasonal Affective Disorder (SAD), 338–40
Selina (rat), 78–80
Shakespeare, William, 328, 329
Shickman, Dora, 278–9
shrews, 326–31, 337
skuas, 321
Sloane, Barney, 68–9
slugs see molluscs
snails see molluscs
Snyder, Gary, 207
Solnit, Rebecca, 96
Solomon, King, 123, 124, 235
solstices, 40, 236
Sparks, T. H., 92
sparrowhawks, 274–5, 278
sparrows, 268–75
    in author's garden, 334
    numbers, 272–5, 278–9, 281–2
    purpose, 278–9

INDEX

species *see* biodiversity
spiders, 123–40
 biology and characteristics, 128–30, 131–3
 and climate change, 135–6
 fear of, 124–6, 128, 137
 in literature, 130–1
 mating, 128–9, 132
 origins, 127
 otherness, 128
 as parents, 132
 place in ecosystem, 136
 travel, 130
 webs and silk, 122, 127, 129–30, 133–4, 135, 138
spoonbills, 267
spring cleaning, 121–2
squirrels, 288–92, 299–307
 food caching, 290
 as pests, 292, 299–307
 red, 288, 291–2, 300–5
Stardust Memories (film), 181
starlings, 53–7, 145, 270, 271
'The Starlings in George Square' (Morgan), 56–7
Statistical Accounts of Scotland, 302–3
Steinbeck, John, 329
Steingraber, Sandra, 270
Stephen, Mark, 29–30
Stevenson, Robert Louis, 94
stoats, 225–6
Stubbs, John Heath, 59–60
Summers-Smith, Dr J. Denis, 273
sun: eclipses, 43
swallows, 228–30, 248
swans, whooper, 48
swifts, 228–9, 313, 332, 337

Taborsky, Michael, 75–6
Tallmadge, John, 149, 150
tanagers, 146–7
Taoism, 31–2, 235
Tel Megiddo, 97
Thomson, Arthur Landsborough, 331
Thoreau, Henry David, 270, 278
thundersnow, 28–9
Tinbergen, Niko, 317–18, 319–20, 323
tits *see* blue tits; great tits
toadflax, 88
Toronto, 147

torpor, 337
Trail, James William Helenus, 98–9, 239, 298–9
Treat, Mary, 138
Trillin, Calvin, 168
twilight, 53
Tyson, Mike, 183

Uria aalge, 109
Uria troile troile, 105, 109

Venice, 119, 189–92
Vincent, Dr Kate, 273
violence
 human infliction of pain on other animals, 212–16
 human vs non-human, 113–17
vivisection, 74–7

Wampanoag people, 93
Watts, Alan, 31–2
waxwings, 33
Weart, Spencer R., 231–2
weasels, 225–6
weather
 British love of discussing, 48–9
 Scottish attitude, 49–52
whales, 311
White, E. B., 130–1
White, Evelyn C., 170–1
White, Gilbert, 91, 328
Whyte, Sandy, 243–4
wilderness concept, 149–51, 170–1; *see also* countryside
Wilson, E. O., 201–3
wind, 221–4
winter
 and mood, 338–40
 preparations for, 335–8
witch hazel, 61
woodpeckers, 229
worms, 214–17

yin and yang, 31–2
Ythan Estuary, 267

Zalasiewicz, Dr Jan, 233
Ziki (crow), 27–8, 42, 250, 312–13, 341
zoonotic diseases, 55–6, 68–9, 190–1

Keep in touch with
Granta Books:

Visit grantabooks.com to discover more.

GRANTA